ANDROGENS

AND ANABOLIC AGENTS

CHEMISTRY AND PHARMACOLOGY

JULIUS A. VIDA

WORCESTER FOUNDATION FOR EXPERIMENTAL BIOLOGY
SHREWSBURY, MASSACHUSETTS

1969

ACADEMIC PRESS   New York and London

ACADEMIC PRESS, INC.
111 Fifth Avenue, New York, New York 10003

United Kingdom Edition published by
ACADEMIC PRESS, INC. (LONDON) LTD.
Berkeley Square House, London W.1

LIBRARY OF CONGRESS CATALOG CARD NUMBER: 68-23482

263906

PRINTED IN THE UNITED STATES OF AMERICA

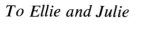
To Ellie and Julie

PREFACE

Many chemists, working in the field of steroids, have been interested in the subject of chemical structure, biological activity relationship. As the number of synthetic steroids increases at an ever-growing rate, more and more theories are put forward to explain their mode of action.

The aim of this work is to review the structure–activity theories in the field of androgens and anabolic agents. It is written primarily for medicinal chemists and pharmacologists, but other research scientists, such as endocrinologists and biochemists, may also find it of value.

The first chapter deals with metabolic factors affecting the biological activities of androgens. In the second chapter the structure–activity theories are reviewed. A new approach to the structure–activity relationship is offered in Chapter 3. The effects of structural and stereochemical changes on biological activities are extensively analyzed in this chapter. A new theory of steroid–receptor interaction concludes this chapter.

In Chapter 4 the therapeutic action of androgens and anabolic agents is examined. In Chapter 5 some 650 androgens and anabolic agents are compiled. The literature is covered through 1967. Some compounds could not be incorporated into the tables owing to the lack of biologic data.

I am very grateful to the late Dr. Gregory Pincus and to Dr. Marcel Gut (Worcester Foundation for Experimental Biology) for their encouragement, and to Dr. Hudson Hoagland (Worcester Foundation for Experimental Biology) and Dr. Endre Balazs (Retina Foundation, Boston) for their help in promoting the publication of this work. I also want to thank Dr. Howard E. Ringold (Worcester Foundation for Experimental Biology; presently with Syntex Research, Palo Alto) for his aid. He not only read the initial outline of this study but also severely criticized it, and his many valuable suggestions helped to guide this work into its present form.

My thanks are due to Mrs. Mina Rano (Worcester Foundation for Experimental Biology) for typing the manuscript, to Mrs. Janet Maglione (The Kendall Company, Theodore Clark Laboratory, Cambridge, Massachusetts) for secretarial assistance and to Mr. William R. Wilber (The Kendall Company, Theodore Clark Laboratory, Cambridge, Mass.) for his help in preparing the index.

Cambridge, Massachusetts   Julius A. Vida*
October, 1968

*Present address: The Kendall Company, Theodore Clark Laboratory, Cambridge, Massachusetts.

CONTENTS

Chapter 4. **Therapeutic Action: Anabolic–Androgenic Ratios**

Chapter 5. **Tables**

ANDROGENS

AND ANABOLIC AGENTS

CHEMISTRY AND PHARMACOLOGY

CHAPTER 1

GENERAL ASPECTS

INTRODUCTION

In order to gain knowledge about the relationship of chemical structure and biological activity, we have investigated many structurally different steroids. In the field of steroids, relatively small changes in the chemical structure may bring about sharp changes in potency. In order to explain drug action in chemical terms the receptor theory [1] was advanced; the theory considers how the drug molecule alters the chemical properties or interferes with the biochemical reactions of living tissues. Several hypotheses have been proposed [2] for drug–receptor interactions, many in the field of steroids [3, 4, 6, 7,* 43, 71, 87, 100, 107]. We will review these theories; in addition, we will study the effect of structural, configurational, and conformational changes upon hormonal activity.

ANDROGENS AND ANABOLIC AGENTS

Androgens are defined as a group of biologically active steroidal compounds which are responsible for the secondary sex characteristics of the male.

Kochakian and Murlin [166] discovered that extracts of male urine injected into dogs not only had an androgenic effect but, in addition caused retention of nitrogen (anabolic effect). Since then, numerous investigations have led to the realization that steroid hormones have many potent effects on the body and minor chemical alterations of the steroid molecule may increase some of these effects without affecting others. Since the anabolic properties of androgens have been confirmed on both physiological and pharmacological grounds [9], many syn-

---

*I am very grateful to Dr. I. E. Bush for an advance copy of his paper.

thetic androgens have been produced with the hope of a useful degree of protein anabolism without androgenic activity.

Anabolic steroids are those compounds which cause nitrogen retention, i.e., positive protein metabolism. In man, treatment with testosterone and related compounds not only brings about nitrogen retention, but potassium, phosphorus, and calcium retention as well. These elements, together with water, are the main components of the intracellular compartments; hence nitrogen retention is essential in the formation of new protoplasm. Since anabolic steroids promote protein synthesis in the muscular system, these drugs find important application in clinical medicine to speed up healing of wounds and to promote muscular regeneration after surgery or wasting diseases. Another clinical application of anabolic steroids is made to counteract the antianabolic or catabolic effects of the extensively used corticoids. Some androgenic and anabolic agents are also able to induce objective remission and prolong life in women with advancing carcinoma of the breast.

Theoretically, for clinical use, anabolic steroids should possess the truly anabolic activity of typical androgens such as testosterone but should lack all androgenic properties, such as virilizing effects. Such compounds have not been reported as yet. In practice, one considers as anabolic steroids those compounds in which anabolic activity prevails and which do not show the virilizing effect of androgens in dosages which are clinically active on protein metabolism.

Chemical modifications of testosterone have led to some synthetic compounds which show a satisfactory dissociation between anabolic and androgenic properties. Many attempts have also been made to find orally active anabolic steroids without androgenic activity since orally active compounds would have a much greater clinical application than intramuscularly administered drugs.

Excellent accounts of the biological effects and clinical application of anabolic steroids, by Krüskemper [5], and of the pharmacology of anabolic steroids, by Overbeek [50], are available. It is the aim of this book to examine the chemical structure–biological activity relationship of androgens and anabolic agents. A new theory of steroid–receptor interaction is also advanced, based on a new approach to the chemical structure–biological activity relationship.

PRODUCTION OF ANDROGENS IN MAN

Hormones are the chemical products of the endocrine glands. They are delivered to the blood stream and transported to target organs. After attachment to a receptor site, the hormones accelerate or decelerate normal biochemical processes. They may be used up (metabolized) in this function or may be released to return into the blood stream. Thereafter the hormones may be metabolized in the blood or may reach a metabolizing organ; the metabolic products are subsequently excreted in the urine.

Testosterone, the natural androgen, is produced by the Leydig cells of the testis under normal conditions. Testosterone is transported in the plasma by means of carrier proteins. This protein binding is of a weak and reversible nature [235]. In the liver, which is the main site of inactivation, and possibly also in the tissues where it exerts its action, testosterone is metabolized. The metabolic products possess reduced androgenic activity compared to testosterone; some metabolic products are inactive as androgens. The metabolic inactivation is carried out by enzymes, such as 17β-dehydrogenase, which is responsible for a metabolic transformation of testosterone, a steroid containing a 17β-hydroxyl group, to 17-ketosteroids. Following such metabolic transformations the steroids are conjugated with glucuronic and sulfuric acids and then excreted in the urine.

The biosynthetic pathways employed by different organs, like testis, ovary, and adrenal cortex, are common to all these organs. Steroids owing their structural differences to their origin of biosynthesis are interconvertible. Pregnenolone is considered a primary steroid on the route to steroid hormone formation, while progesterone is an important intermediary steroid in the biosynthesis. Progesterone, which is produced in the ovary, and dehydroepiandrosterone, which is a product of adrenal cortex, are considered precursors of androgens in the biosynthetic pathways. Thus endocrine differences between testis, ovary, and adrenal cortex are quantitative rather than qualitative.

The composition of the excreted urinary 17-ketosteroids also reveals a close similarity between the hormones of different origin: the testis accounts for 30% while the adrenal cortex contributes the remaining 70% of the total urinary 17-ketosteroids [383]. Androsterone, epiaandrosterone, and 5β-androsterone (etiocholanolone) are the main urinary metabolities of testosterone, and dehydroepiandrosterone is the major urinary 17-ketosteroid derived from the adrenal cortex.

The intermediates employed in the biosynthesis of androgens belong to two structurally different chemical classes, α,β-unsaturated ketones (progesterone) and β,γ-unsaturated alcohols (pregnenolone and dehydroepiandrosterone). The biosynthetic pathways in the production of androgens are also different [390].

A recent review of modes and sites of action of steroid hormones has been given by Engel [391].

METABOLIC FACTORS

Before considering the structural requirements of androgens and anabolic agents, one must take into account the metabolic factors affecting the biological activities of steroid hormones. The question can be raised whether the steroids administered as drugs still possess the original structure of the drug at the receptor site or whether structural changes have taken place in the course of the delivery to the receptor site. In other words, does the receptor see the same structure that we see at the time of administration? Is the transport of the compound to the receptor site in the organism fast enough and is the compound protected from metabolic inactivation long enough to permit delivery of the active species to the end organ? What is the half-life of the steroid in the body? These questions become especially important in the view of the finding [167] that the truly anabolic effect of testosterone (i.e., a definite positive change in nitrogen balance) begins only at a time when all but traces of the effective dose of the steroid have been metabolized and excreted by the body.

Dorfman and Ungar have given an account of the metabolism of steroid hormones [10], therefore only a few comments will be made here. The comparative relative activities of testosterone metabolities on the seminal vesicle and ventral prostate of immature male rats and capon's comb are summarized in the accompanying table.

It can be seen that androstane-17β-ol-3-one (dihydrotestosterone) is 2 to 2½ times as active as testosterone on the seminal vesicles and ventral prostate indices, but the rest of the compounds are all less active than testosterone. Thus metabolic inactivation is a very important factor to be considered.

There are several approaches to the study of metabolism. First, isolation of steroid metabolites from various steroid-forming tissues (testes, ovaries, placentae, adrenal glands) makes it possible to evaluate the usefulness of a particular endocrine gland in the production of steroids. Secondly, *in vitro* experiments are conducted by incubation with tissue preparations followed by isolation of the metabolities of steroids. Thirdly, *in vivo* experiments are carried out by isolation of steroids from biological fluids, mainly from animal and human blood and urine. This last method includes evaluation of the rates of secretion and metabolism of

Metabolites of testosterone	Testosterone = 100% Relative activity (%)		
	Seminal vesicles [8]	Capon's comb [8]	Ventral prostate
Testosterone	100	100	100
Androstane-17β-ol-3-one	200	75	260
Androstane-3α,17β-diol	33	75	24
Androstane-3β,17β-diol	10	2	3
Δ^4-Androstene-3,17-dione	20	12	39
Androstane-3,17-dione	14	12	33
Δ^5-Androstene-3,17β-diol	14	3	21
Androstane-3α-ol-17-one (Androsterone)	10	10	53
Androstane-3β-ol-17-one (Epiandrosterone)	3	2	2
Δ^5-Androstene-3,17-dione	7	12	–
Δ^5-Androstene-3β-ol-17-one (Dehydroepiandrosterone)	3	16	34

steroid hormones by the application of isotope tracer techniques [318–321]. Lieberman and co-workers undertook urinary metabolic studies [304, 325] by administration of isotopically labeled steroids followed by the determination of specific activities of urinary metabolites derived only from the administered substance. Mathematical analyses were performed by compartmentalization of the metabolism [385]. One- or two-compartment systems are most frequently used [304, 306, 384, 386, 387]. A compartment is defined as a particular species (e.g., dehydroepiandrosterone sulfate) in a particular space (e.g., peripheral organ). It is assumed that a species entering a compartment mixes immediately with the whole compartment. Two factors were responsible for the institution of compartmental analysis:

(1) Consider three structurally similar C_{19} steroids — testosterone, androstenedione, and dehydroepiandrosterone. Each of these compounds is secreted into the blood stream by glands, and after provoking its biochemical effect, is metabolized and excreted. If the only source of the plasma pool were provided by the glandular secretion of the hormone, then the secretion rate would be used to calculate the plasma concentration. However, the picture in the case of the three C_{19} steroids is complicated by the fact that the three compounds are peripherally interconvertible. Since both glandular secretion and peripheral conversion of precursors contribute to the plasma pool, the production rate is used to

calculate the plasma concentration. The production rate is expressed as the sum of the secretion rate of the steroids entering the circulation directly and the steroids entering irreversibly from the outer anatomic pool [318].

(2) It has been shown that hormones are not exclusive products of the glands but are also formed in metabolizing organs. These hormones, however, contrary to the classical hormones, are not secreted into the blood. It has been established that testosterone formed in the liver from androstenedione, dehydroepiandrosterone, and dehydroepiandrosterone sulfate does not enter the blood [305, 323, 388]. It has also been established that secreted and metabolically produced testosterone do not have the same metabolism [169, 311].

Compartmental analysis has been proved to be useful in dealing with the complex situation of metabolism.

In addition to estimation on the basis of urinary metabolic studies, blood production rates of steroids can be estimated, after administration of isotopically labeled steroids, from the specific activity of the unchanged steroids in the blood. This method was extensively used by Tait and co-workers [322, 324]. In 1966 a critical survey of steroid hormone metabolism was given by Baulieu [385].

METABOLIC INACTIVATION

The main factors responsible for metabolic inactivations of the androgens are listed below; they are discussed in the following sections.

1. Metabolism of the 4,5-double bond
2. Metabolism of ketone and secondary hydroxyl groups
3. Conjugation of hydroxyl groups
4. Hydroxylation and epoxidation reactions
5. Aromatization of androgens
6. Other metabolic factors

METABOLISM OF THE 4,5-DOUBLE BOND

The natural androgen testosterone and a large number of other androgens and anabolic agents possess the 4,5-double bond, while the majority of metabolic products lack this unsaturation. The hypothesis was put forward that the reduction of the 4,5-double bond in the liver may represent the rate-limiting step in the inactivation of these hormones [168, 221]. Upon hydrogenation of the double bond, the 5-carbon becomes asymmetric and therefore two possible isomers could result, the 5α-isomer (trans

A/B ring junction) and the 5β-isomer (cis A/B ring junction). Catalytic reduction of simple 4,5-unsaturated compounds (without neighboring ketone or alcohol functions) in general gives a mixture in which the 5β-isomer (A/B cis product) predominates. In the case of Δ^4-cholestene the course of reduction of the 4,5-double bond is dependent upon the medium; in a neutral medium Δ^4-cholestene is reduced to the 5β-isomer (A/B cis product) and in an acidic medium to the 5α-isomer (A/B trans product) [330]. One explanation was advanced [329] for the usual preponderance of β-isomers in catalytic reductions on the basis of conformational analysis [331,332] and another [333] on the basis of catalyst hindrance theory [334]. The direction of attack can be controlled by steric factors imposed by nearby substituents; catalytic reduction of 3β-hydroxycholest-4-ene, an allylic alcohol, gave a different composition of the mixture of stereoisomers [335]. Catalytic reduction of the α,β-unsaturated ketone, cholest-4-en-3-one, gives a mixture of saturated 5α (A/B trans) and 5β (A/B cis) ketones [336]; the proportion of products varies with the conditions and with the structure of remote parts of the molecule [337]. A hydroxyl group at the 11β-position leads to exclusive production of 5α (A/B trans) saturated ketones [338], while 11α-hydroxyl groups lead to exclusive formation of saturated 5β (A/B cis) ketones in the course of catalytic hydrogenation of Δ^4–3 ketone compounds [339]. In addition to catalytic reduction, the α,β-unsaturated ketone system can be reduced to saturated ketone by metal-ammonia reduction [340]. This kind of reduction will give rise to that product which will be the more stable of the two isomers (cis or trans) having the newly introduced hydrogen axial to the ketone ring [341].

Considering the courses of the chemical reductions of the Δ^4–3-ketone system, it is not surprising to find that the enzymatic metabolic reductions are very similar in behavior to the chemical reductions. Thus, reduction of the 4,5-double bond gives rise to a mixture of 5α- and 5β-steroids and the ratio depends on the nature of substituents in the molecule. Without an 11-oxygen function the ratio of 5β- to 5α-dihydro derivatives was found to be 1:1 [342] to 2:1 [301]. In the presence of an 11-oxygen function (11-hydroxy [343,344] and 11-keto [345]) the ratio of 5α- to 5β-reduced metabolites was 4:1 or 5:1.

Since the 5β-dihydro derivatives, compounds in which rings A and B possess a cis junction, are inactive as androgens or anabolic agents, the ratio of products formed in the reduction becomes important.

METABOLISM OF KETONE AND SECONDARY HYDROXYL GROUPS

Since reduction–oxidation processes are very common in all living tissues, all ketone and secondary hydroxyl groups are susceptible to redox conversions. This factor is extremely important with respect to the 17-ketone–17-hydroxyl group transformation. Testosterone, which possesses a 17β-hydroxyl group, is known as the most potent natural androgen and androstenedione, which possesses a 17-ketone group, is a very weak one; therefore the protection of the 17β-hydroxyl group against oxidative metabolic inactivation becomes a very important consideration. The reversible reaction androstenedione ⇌ testosterone has been demonstrated in the liver, the kidney, and other tissues [171–173]. It has been assumed [174] that both testosterone and dehydroepiandrosterone contribute to a plasma androstenedione pool. Androstenedione was considered the precursor of further metabolites. Accordingly the 17β-hydroxyl group of testosterone is first oxidized to the 17-ketone group and then it is further reduced to other metabolites, i.e., it follows the "ketonic metabolic pathway." Baulieu [169] has demonstrated that a direct "hydroxyl pathway" also exists, and it is not necessary to first

Androstenedione   Testosterone

Ketonic Pathway   Hydroxylic Pathway

Epiandrosterone   Androsterone   5 β- Androsterone

Androstanol   5 β- Androstanol

oxidize the 17β-hydroxyl group for other metabolic transformations to take place.

Androsterone and 5β-androsterone, which are the major metabolites, were thought to be uniquely derived from the plasma androstenedione pool. Korenman et al. [170, 388] demonstrated by the double isotope tracer method using carbon-14-labeled testosterone and tritium-labeled androstenedione that neither androsterone nor 5β-androsterone is a unique metabolite of a plasma androstenedione pool. A unique steroid metabolite has been defined by Dorfman [326] as a steroid persisting or formed during metabolism which can be related to one and only one tissue steroid. Korenman et al. obtained different tritium/carbon-14 ratios for androsterone and 5β-androsterone, suggesting that other pathways of testosterone and androstenedione metabolism also exist. These other pathways, possibly metabolism of testosterone and androstenedione by peripheral tissue, may be responsible for the relative enrichment of either the 5α- or the 5β-isomer of the urinary metabolites [305].

Furthermore Tait and co-workers [322] pointed out the marked difference in the urinary and blood production rates of testosterone obtained in women after injection of radioactive testosterone. He concluded that steroids produced from dehydroepiandrosterone contribute little to the blood production rate of androstenedione and testosterone in normal subjects [403]. According to Tait [324], all the blood production rate of androstenedione in the female and testosterone in the male is due to the same secreted steroid, while the blood production rate of testosterone in the female and androstenedione in the male is due about one-half to the same secreted steroid and one-half to converted precursor. The normal male secretes a ratio of testosterone to androstenedione of about 10 : 1 and the normal female secretes a ratio of androstenedione to testosterone of about 25 : 1.

We have already considered the metabolism of the 4,5-double bond, including the hypothesis that the reduction of the 4,5-double bond may represent the rate-limiting step in the inactivation of hormones possessing the Δ^4–3-ketone system. After the reduction of the 4,5-double bond, the 3-ketone function is also reduced, giving rise to a new asymmetric center at carbon-3. A correlation between the stereochemical course of the chemical reduction and the conformations of alcohols has been established by Barton [331, 346]. The reduction of 3-ketones with sodium in alcohol yields predominantly the thermodynamically more stable equatorial alcohol. Reductions by alkali metals and proton donors proceeding through carbanions also give predominantly the thermodynamically most stable product, providing the carbanions are sufficiently long-lived to exercise their stereochemical preference [347]. The von Auwers–Skita hydrogenation rule [348] was applied to the stereochemical course of catalytic reduction of 3-ketones and later modified by Wicker [349, 350]. In strongly acidic solution, rapid catalytic hydrogenation yields predominantly the less stable axial alcohol. It was postulated by Dauben [351] that two factors control the stereochemistry of the products obtained from the reduction of 3-ketones with complex hydrides. In the mixture of isomeric alcohols, the equatorial one will predominate due to the ease of formation of the initial complex between the hydride and the relatively unhindered 3-ketone group. In this case, the reaction course is influenced primarily by thermodynamic factors; the reduction will follow the "product development control." The composition of the mixture varies depending on the steric requirement of the different metal hydrides used in the reduction [351, 357]. In contrast, the reduction of 2-ketones and 4-ketones will follow the "steric approach control." In these hindered systems, the axial approach of hydride is strongly inhibited by an axial methyl group; thus the hydride is forced to approach equatorially and the product will be the axial alcohol. Other mechanisms for the reduction with complex hydrides have also been offered [352–355]. H. C. Brown [357] pointed out that in cases where steric hindrance becomes a predominating factor, steric hindrance to the departure of the leaving group introduces a new factor which might be termed "steric departure control." An interpretation was offered recently by postulation of a flexible form of the ketone in the reaction [393].

The reduction of the 3-keto group of androgens to both 3α- and 3β-hydroxyl groups has been demonstrated in both *in vivo* and *in vitro*. Metabolic studies demonstrated a sex difference in the reduction by rat liver homogenates of the 3-ketone group of 5α-androstane-3,17-dione and androst-4-en-3,17-dione. When these steroids were incubated [356] with a rat liver homogenate prepared from male rats, the ratio of 3β-ol (epiandrosterone) to the 3α-ol (androsterone) varied from 2.3 to 3.5 while a similar preparation from female rats gave ratios varying from 0.12 to 0.16. *In vivo* evaluation of the ratio of 3α- and 3β-hydroxy epimers is made more difficult since the metabolic products may arise from C_{21} steroids as well as C_{19} steroids. Also, there is no unique metabolite derived exclusively from androst-4-en-3,17-dione since the metabolic products may arise from transformations of testosterone or dehydro-epiandrosterone as well [170].

CONJUGATION OF HYDROXYL GROUPS

Reduction of the 3-ketone group can give rise to 3α- or 3β-hydroxy compounds, with the former dominating in men. Of the two epimers, the 3α-hydroxy compounds are usually the more potent androgens and anabolic agents. A contrary example was given by Dorfman [241];

$3\beta,17\beta$-dihydroxyandrost-4-ene in mice, rats, and chicks is about as androgenic as testosterone. In the reduction of the enone system of a Δ^4-3-ketone group the rate-determining step is saturation of the double bond. After saturation of the double bond, the reduction proceeds almost quantitatively to the hydroxy compounds regardless of the redox potential. This is probably due to the irreversible removal of the hydroxy compound from the ketonic compound–hydroxylic compound equilibrium by formation of so-called conjugates, e.g., esters of sulfuric or glucuronic acid. The glucuronide formation [327] may approximate in vivo an irreversible process and efficient route of inactivation. Metabolic studies of dehydroepiandrosterone glucuronide showed [385] that when isotopically labeled dehydroepiandrosterone and dehydroepiandrosterone glucuronide are injected intravenously, dehydroepiandrosterone sulfate is formed from both (in a ratio of 5:4 favoring formation from the free compound). This indicates the existence of a transconjugation mechanism without the need for free dehydroepiandrosterone intermediate. On the other hand, the conjugated dehydroepiandrosterone glucuronide undergoes further metabolic transformations [312], such as androsterone formation, to a very limited extent (10% of that of free dehydroepiandrosterone). This shows that dehydroepiandrosterone glucuronide has a greater tendency to form dehydroepiandrosterone sulfate than does free dehydroepiandrosterone, the latter having a greater tendency to undergo further metabolism rather than conjugation with sulfuric acid [311, 385].

Formation of sulfates [328] is relatively reversible. This may be due to the fact that the metabolic clearance rate [318] of sulfates is relatively low [389] and their renal excretion inefficient [4]. The conversion of dehydroepiandrosterone sulfate in vivo to androsterone and etiocholanolone glucuronide was demonstrated by Lieberman and co-workers, who administered isotopically labeled dehydroepiandrosterone sulfate and isolated labeled androsterone and etiocholanolone from glucuronidase-hydrolyzed urine [303]. Since in another study by Lieberman [174], the transconjugation from dehydroepiandrosterone sulfate to dehydroepiandrosterone glucuronide without free dehydroepiandrosterone intermediate was declared to be improbable, in vivo equilibrium between dehydroepiandrosterone sulfate and dehydroepiandrosterone seems probable [303].

According to the classical view of metabolism the hormones are synthesized as free steroids in endocrine tissues and prepared for excretion in urine by peripheral metabolism and conjugation. This view had to be modified upon the isolation of dehydroepiandrosterone sulfate from adrenal tumor [307]. Thus dehydroepiandrosterone sulfate, a steroid conjugate, was shown to be secreted by the adrenal tissue. Isotopic methods also pointed in the same direction. Lieberman et al. [304], using the isotopic dilution method, found that after administration of tritium-labeled dehydroepiandrosterone to normal subjects, the specific activity of dehydroepiandrosterone excreted as glucuronide was higher than that of dehydroepiandrosterone excreted as sulfate, indicating that dehydroepiandrosterone could be secreted. The double isotope tracer method using carbon-14-labeled dehydroepiandrosterone and tritium-labeled dehydroepiandrosterone sulfate was applied [306] to estimate the secretory rates of both dehydroepiandrosterone and dehydroepiandrosterone sulfate and the results confirmed that dehydroepiandrosterone sulfate is one of the major steroids secreted by human adrenals.

However, the validity of the model used by Gurpide *et al.* [174, 175, 176, 306] for the determination of the secretion and interconversion of dehydroepiandrosterone, dehydroepiandrosterone sulfate, androstenedione, and testosterone was criticized by Baulieu *et al.* [404] and Migeon *et al.* [405].

Baulieu isolated dehydroepiandrosterone sulfate from two adrenal tumors [307] and found that the concentration of dehydroepiandrosterone sulfate in adrenal venous blood was higher than that present in peripheral blood, implying that dehydroepiandrosterone sulfate was synthesized by the gland [308, 308a]. Although dehydroepiandrosterone sulfate is a secreted as well as a metabolic steroid conjugate, it is not an end product in the course of metabolism; it, too, as already mentioned, undergoes further metabolism [312].

The principal point of conjugation (ester formation) is at the hydroxyl group attached to C-3. However, if ester formation at this position is not possible and there are other hydroxyl groups present in the molecule, these other hydroxyl groups also may become attached to glucuronic or sulfuric acid to give rise to steroid conjugates.

HYDROXYLATION AND EPOXIDATION REACTIONS

Hydroxylation reactions lead to metabolites which are usually less active than the precursors. Hydroxylation reactions may take place at a position activated in the chemical sense, such as an allylic methylene group or a position adjacent to or vinylogous to a carbonyl function. Hydroxylations of saturated carbon atoms that are inactivated in any classical sense are also important. Bloom [358] made a discovery which provided the basis for his proposed mechanism for oxidative attacks on steroids, including hydroxylations and epoxidations. According to Bloom "a microorganism capable of introducing an axial hydroxyl function at C_n of a saturated steroid also effected the introduction of an epoxide grouping axial at C_n in the corresponding unsaturated substrate." Equatorial hydroxylases did not effect a similar conversion. Bloom [359] assumed that hydroxylations proceed by electrophilic attack. Molecular orbital considerations reveal [360, 361] that in a cyclohexene system the unsaturated carbon atoms possess sp^2-hybridized orbitals with the maximum π-electron density extending directly above and below the plane of the molecule. Maximum overlap of π-electrons for the formation of a new bond is expected to occur with axial attack. In the saturated cyclohexane system the carbon atoms possess sp^3-hybridized orbitals and the maximum electron density of a carbon–hydrogen bond is concentrated along the σ-type bond. The spatial requirement of the electron bond along the axial — not the equatorial — carbon–hydrogen bond is similar to that of the unsaturated carbon–carbon bond. Therefore, as outlined by Bloom [359], in the appropriate enzyme–substrate complex the spatial relationship required for reaction (hydroxylation) between an axial hydroxylase and an axial hydrogen bond of a saturated steroid is also similar to the spatial relationship required for reaction (epoxidation) between the axial hydroxylase and the π-electrons of the unsaturated steroid. A reasonable degree of structural specificity in the oxidizing enzyme system would preclude occurrence of the epoxidation reaction with equatorial hydroxylases since equatorial bonds extend outward in the plane of the cyclohexane ring.

The electrophilic nature of peroxy acids commonly used in preparation of epoxides is recognized in all of the mechanisms postulated for the epoxidation reaction [362, 363, 373]. Enzymatic hydroxylations of saturated carbon atoms that are unactivated in the chemical sense proceed with retention of configuration [314–317]. Implications are that enzymatic hydroxylating agents also possess an electrophilic character, on the basis of following consideration. Comparison of the stereochemical course of nucleophilic and electrophilic substitution reactions reveals that with the exception of nucleophilic internal substitution ($S_N i$) of alcohols with a limited number of reagents, which results in substitution with predominant retention of configuration, the nucleophilic substitution reaction does not proceed with retention of configuration [376]. On the other hand, electrophilic substitution reactions at saturated carbon often proceed with retention of configuration, especially in solvents of low dissociating power [374, 375]. Enzymatic hydroxylation procedures utilize molecular oxygen [364, 365] and require reduced nicotinamide adenine nucleotides [313, 366].

Ringold put forward [369] a hypothesis for the enzymatic hydroxylation reactions which take place at an activated position. These reactions at a position adjacent or vinylogous to a carbonyl function may take place via the enol intermediates. Assuming an electrophilic hydroxylating species, a high electron density provided by the enolic form at the position of hydroxylating attack may greatly accelerate the rate of reaction. This hypothesis is in line with the general concept that through enol or enolate ion formation, a carbonyl group activates adjacent carbon–hydrogen bonds toward electrophilic substitution [367, 368]. Since maximum overlap of π-electrons for the formation of new bond is provided by an axial approach of the hydroxylating agent, axial 2β-, 6β-, 10β-, and 17α-hydroxylations are favored over equatorial 2α-, 6α-, 10α-, and 17β-hydroxylations. Since 17α-hydroxyprogesterone is considered an important metabolite in the biosynthetic pathways of steroids, 17α-hydroxylation is a most significant reaction. The mechanism of this reaction was discussed by Hayano [370].

AROMATIZATION OF ANDROGENS

The conversion of radioactive testosterone to estradiol has been demonstrated with mammalian tissue [225] and accepted as a natural step in estrogen formation. The conversion of 19-hydroxy-Δ^4-androstene-3, 17-dione to estrone by endocrine tissue [226] suggested that the bioconversion of androgens to estrogens proceeds via a C-19-hydroxylated intermediate. Simple removal of the angular methyl group (as in 19-nortestosterone), or use of 1,4-dien-3-one with the angular methyl group resulted in reduced relative substrate activity in steroid aromatization [227]. This suggested that aromatization proceeds via C-19 hydroxylation followed by simultaneous or rapidly sequential direct introduction of a double bond (1,2-unsaturation) and removal of the angular methyl group [371, 372]. Substitutions in the molecule at C-11, C-16, and C-17 all diminish or abolish activity, but 11α-hydroxytestosterone and 6β-hydroxy-4-androstene-3,17-dione are apparently converted to substituted estrogens. The unsaturated compound, 1-androstene-3,17-dione is also active in spite of the lack of direct evidence for mammalian enzyme systems capable of inserting a double bond in the C-4 position [227, 377]. Hydroxylation at C-1 and C-2 prior to other changes in the molecule results in substantial loss of activity and there is also evidence that hydroxylation and dehydration are probably not involved in double bond formation [227]. Ester formation at the C-17 hydroxyl group does not protect against aromatization; esters of testosterone and nortestosterone gave rise to approximately the same increase in urinary estrogens as the free steroids [233]. Alkylation in the 17α-position causes a marked decrease in steroid aromatization, but offers no complete protection, i.e., 17α-methyltestosterone has a 44% activity in aromatization compared to 100% testosterone activity. The ring-A saturated steroid, 17β-hydroxy-androstan-3-one does not aromatize [227]. Chlorine substitution in the 4-position also inactivates the steroid towards aromatization, i.e., 4-chloro-17β-hydroxyandrost-4-en-3-one 17β-p-chlorophenoxyacetate and

4-chloro-17α-methyl-17β-hydroxyandrosta-1,4-dien-3-one administration does not cause estrogen excretion [234]. The 1-methyl-substituted steroid, 1-methyl-17β-hydroxyandrost-1-en-3-one does not aromatize; the 1-methyl group inhibits the initial attack of 19-hydroxylase, which appears to be the first step in the aromatization [228].

OTHER METABOLIC FACTORS

Metabolic studies have shown that after intravenous administration of carbon-14-labeled testosterone the radioactivity was excreted almost quantitatively in the urine in 48 hr [310]. However, the sum of all the known radioactive metabolites is far below 100%. Intravenous administration of labeled metabolites, on the other hand, made it possible to calculate the conversion rate of metabolites on the basis of the specific activity of the urinary metabolites. The calculated production rates of known metabolites showed that testosterone is almost quantitatively transformed to known metabolites, then the formed metabolites in turn are converted in various degrees (less than 100%) to other urinary metabolites. The metabolites do not circulate as free compounds in the blood; they are present as conjugates. Therefore, further metabolic studies were carried out using conjugated steroids [311]. Tritium-labeled testosterone and carbon-14-labeled testosterone glucuronide were intravenously injected. Comparison of the urinary metabolites showed that while tritium-labeled testosterone affords the normal conversion to tritiated 5α (epiandrosterone, androsterone, and androstanediol) and 5β (5β-androsterone, 5β-androstanediol) metabolites, the percent conversion of carbon-14-labeled testosterone glucuronide to carbon-14-labeled 5β (5β-androsterone and 5β-androstanediol) metabolites is relatively higher. However, most striking is the fact that carbon-14-labeled testosterone glucuronide gives rise to practically no carbon-14-labeled 5α (epiandrosterone, androsterone, 5α-androstanediol) metabolites. This seems to indicate that testosterone glucuronide is not split to give testosterone and that an unknown pathway from testosterone glucuronide to the 5β-steroids could exist [311].

Metabolic studies carried out with isotopically labeled dehydroepiandrosterone sulfate [312] showed that this conjugated steroid may follow an "indirect" metabolic pathway initiated by the hydrolysis of the sulfate group. In the course of the metabolism the conjugated steroid thus becomes a free steroid first and the free steroid may undergo further metabolism. On the other hand, dehydroepiandrosterone sulfate may follow a "direct" metabolic pathway without a break of the ester group.

These studies were carried out by Baulieu and co-workers and are summarized in an excellent review [312].

THE EFFECT OF STRUCTURAL MODIFICATION ON THE METABOLISM

REMOVAL OF THE 19-METHYL GROUP

A very important class of androgenic and anabolic compounds is made up of 19-nortestosterone derivatives. It has been shown that 19-nortestosterone is metabolized in a similar way to testosterone itself [177]. The two main metabolites isolated from the urine of a patient after administration of 19-nortestosterone were 19-norandrosterone (3α-hydroxy-5α-estran-17-one) and 19-noretiocholan-3α-ol-17-one (3α-hydroxy-5β-estran-17-one). The ratio of 5α to 5β,17-ketosteroid metabolites of 19-nortestosterone is similar to the ratio of 5α to 5β,17-ketosteroid metabolites of testosterone. The conclusion was drawn [177] that the absence of the angular methyl group exerts little or no influence upon either the dehydrogenation of the 17β-hydroxyl group or the stereochemical course of the reduction of the Δ^4–3-ketone group. The increased excretion of estrone after administration of 19-nortestosterone is of the same order of magnitude as that observed after the administration of testosterone. Later it was found that the relative substrate activity in steroid aromatization, measured by conversion of androgens to estrogens by the human placenta, was 20% for 19-nortestosterone as compared to 100% for testosterone [227].

PROTECTION OF THE 17β-HYDROXYL GROUP

Another important class of androgenic and anabolic compounds was synthetized with the direct purpose of avoiding metabolic inactivation. Introduction of a 17α-alkyl substituent into the testosterone molecule changes the secondary 17β-hydroxyl group to a tertiary hydroxyl group. Thus metabolic inactivation through oxidation of the 17β-hydroxyl group to a 17-keto group is no longer possible. The 17α-alkyl testosterone derivatives owe their oral activity to the protection of the 17β-hydroxyl group.

In addition to the protective effect of 17α-alkylation, methyl substitution in positions 1, 2, and 6 and the position of unsaturation in ring A also have an important influence on the relative activity of 17β-hydroxydehydrogenase [5]. The relative activities of 17β-hydroxy-5α-androstan-

3-one (11%), 17β-hydroxy-5α-androst-1-en-3-one (18%), 17α-methyl-testosterone (< 10%), 1α-methyl-17β-hydroxy-5α-androstan-3-one (< 10%), 1β-methyl-17β-hydroxy-5α-androstan-3-one (< 10%), 1-methyl-17β-hydroxy-5α-androst-1-en-3-one (< 10%), and 2-methyl-17β-hydroxy-5α-androst-1-en-3-one (< 10%) compared to testosterone (100%) *in vitro* show a pronounced drop in the oxidation of the 17β-hydroxyl group by DNP-specific 17β-hydroxy(testosterone)dehydrogenase prepared from guinea pig liver. Intermediate values in the relative activities were obtained in the same experiment [5] by 1α-methyl-17β-hydroxy-androst-4-en-3-one (47%), 2α-methyl-17β-hydroxyandrost-4-en-3-one (54%), 6α-methyl-17β-hydroxyandrost-4-en-3-one (63%), and 6β-methyl-17β-hydroxyandrost-4-en-3-one (58%). However, in contrast, *in vivo* experiments showed a great increase in the excretion of 17-ketosteroids upon administration of 17β-hydroxy-5α-androstan-3-one (dihydrotestosterone) [237] and 19-nortestosterone [5]. It was also reported [239] that 4-chloro substitution of testosterone also decreases the rate of oxidation of the 17β-hydroxyl group.

Reports have also appeared on the protracted action of 17β-hydroxy-5α-androst-2-ene (A-2)* containing a free 17β-hydroxyl group [399, 400]. The prolonged action is attributed to the specific structural features of the molecule which cause a slow metabolic transformation of its 17β-hydroxyl group to form an inactive 17-keto derivative. In addition, a slower resorption from subcutaneous depot could be another reason [400] for the prolonged action of 17β-hydroxy-5α-androst-2-ene (A-2). In another case, metabolic studies of 1-methyl-17β-hydroxy-5α-androst-1-en-3-one (A-86) showed [275] that the steroid predominantly excreted is the unchanged compound, i.e., the oxidation of the 17β-hydroxyl group did not take place to a large extent. This compound (A-86) was shown to be an orally effective anabolic agent [100, 270, 300]. The results of an *in vitro* experiment on the metabolism of structurally different androstane analogs were also reported by Krüskemper and Breuer [401].

PROTECTION OF THE Δ^4–3-KETONE SYSTEM

Structural modifications were carried out in some classes of steroids in order to block another known route of metabolic inactivation, namely, the reduction of the Δ^4–3-ketone system. The structural changes, such as 2α-methylation or 6α-methylation were aimed to slow down the rate of

---

*The serial number (e.g., A-2) given in parentheses refers to Tables I–V. The tables and directions for their use comprise Chapter 5.

reductions in ring A. The enhanced corticoid activity of 2α-methyl-cortisol was attributed [224] to the apparent inhibition of the reduction of the Δ^4-3-ketone system. This suggestion was confirmed by the large number and amount of steroids retaining the Δ^4-3-ketone system among the metabolites of 2α-methyl steroids [223]. In another class of steroid hormones, 6α-methyl-17α-hydroxyprogesterone acetate has been shown [229] to exhibit a prolonged duration of anti-inflammatory, glycogenic, and pituitary-inhibiting activity in the rat compared to progesterone. Since the metabolic inactivation of 6α-methyl-17α-hydroxyprogesterone acetate by rat liver enzyme is slower than that of progesterone, indications are that the 6α-methyl group protects the Δ^4-3-ketone system from fast reduction. It is interesting to note that in the 5α-androstane series, intro-duction of a 6α-methyl substituent does not lead to an increased andro-genic or anabolic activity [40].

In the progestational series, it was found [232] that of the isomeric C-6 halogen-substituted 17α-acetoxyprogesterone derivatives the active species is the α-isomer (also the thermodynamically more stable isomer). The 6α-halogen-substituted compounds which possess only a 6β-hydrogen are not reduced enzymatically by simple addition of hydrogen at C-4 and C-5 of the Δ^4-3-keto system, but through the enol form in-stead, and formation of the enol proceeds preferentially by abstraction of a 6α-proton. Thus the 6α-halogen derivatives are reduced (metabolic inactivation) at a much slower rate and therefore are more active [3].

Electronic factors may also play an important role in affecting the Δ^4-3-ketone system. For example, it was found [240] that 17β-hydroxy-6β-fluoroandrost-4-en-3-one is reduced by male rat liver supernatant fractions, containing reductase, to a mixture of 3α and 3β,17β-dihydroxy-6β-fluoroandrost-4-ene. The allylic alcohol formation was attributed to the fact that the strongly electron-withdrawing 6β-fluoro substituent brings about polarization of the enone system, which causes the ketone group to resemble a saturated ketone, rather than a conjugated ketone. Therefore only the ketone is reduced [240].

OTHER CHANGES

The metabolism of 17β-hydroxy-1-methyl-Δ^1-androstene-3-one is markedly different from that of testosterone. In addition to the large amount of unchanged compound, small amounts of 1-methyl-Δ^1-andro-stene-3,17-dione and 1α-methyl-androstane-3,17-dione also are ex-creted in the urine, but the 1-methylated androsterone or etiocholane derivatives are not found [275].

The introduction of 7α-methyl substituents into testosterones increases the androgenic and myotrophic activity in every case, while 7α-methylation of Δ^4-androstenedione produces a decrease in androgenicity and has no effect on the anabolic activity. 7α-Methylation of the 19-nortestosterones, even 19-nor-Δ^4-androstenedione, leads to a spectacular increase in androgenic and anabolic activity, averaging about tenfold. It was suggested [35] that the 7α-methyl groups protect the 19-nor configuration to permit delivery of the highly active 19-nor steroid to the end organ.

ORAL ACTIVATION OF STEROIDS

Oral activation of androgens and anabolic agents has usually been promoted by the introduction of a 17α-alkyl substituent into the molecule. The presence of the alkyl group affects the metabolic course of the steroid by preventing the excretion of the steroid as 17-ketosteroid [236]. The presence of the alkyl group also is responsible for undesirable side effects including liver tissue damage. 17α-Alkyl-substituted steroids were reported to cause reversible, intrahepatic obstructive jaundice due to cholestasis [295–298]. The disturbing effects of anabolic steroids on the liver function are tested by sulfobromophthalein retention, serum bilirubin, serum lactic dehydrogenase, and total creatinine chromogen assays [299]. Blood coagulation factors [300] are also used as parameters. 17α-Alkylated steroids produce an increase in sulfobromophthalein retention by interfering with the excretion of sulfobromophthalein (probably in the "conjugate" form). Several 17α-alkylated steroids increase the serum lactic dehydrogenase level. No significant change in serum bilirubin occurred at dose levels which caused a change in the other three parameters. However in patients who became jaundiced elevation of serum bilirubin was found [299]. The use of anabolic steroids resulted in increased creatinine chromogen (creatine plus creatinine) excretion. Since under ordinary circumstances a steady state exists, the rate of endogenous creatine synthesis and the excretion of creatinine chromogen remains constant. The increased creatinine chromogen excretion due to the use of anabolic steroid indicates an increased creatine synthesis.

It is interesting to point out that 1-methyl-17β-hydroxy-5α-androst-1-en-3-one (Methenolone, A-86), an orally active anabolic agent lacking the 17α-alkyl substituent, also caused a slight increase in sulfobromophthalein retention and an increase in the blood coagulation factors V and X [300]. Here the effects are, however, very small. An important class of androgenic and anabolic compounds is made up of 17β-hydroxy esters (acetates, propionates, n-decanoates, phenylpropionates, etc.). The formation of esters is not only effective as a means of protection against metabolic inactivation [402], but also brings about a protracted biological effect. This effect is generally believed to be due to slowing down of the transport of the compound in the organism [178]. This then results in a more prolonged effect on the receptors and a longer lasting weight increase response in the individual organs [382]. Subsequently 17β-hydroxy ethers were introduced [179]. The prolonged action displayed by some ether derivatives is likely to be due to a reduced rate of of elimination from the body. Also the hypothesis was advanced that etherification of a steroid may cause it to be differently conjugated in the body and to exhibit different degrees of affinity for receptors.

In addition to the rate of excretion of the steroids from the body and the rate of metabolic inactivation, two other important factors contribute to the distribution of steroids in the body. One is the solubility property, which is the result of polarity of the steroid molecule as a whole, i.e., the presence or absence of polar groups may render the steroid hydrophilic or hydrophobic. It is remarkable that esterification of the 17β-hydroxyl group with glycine or succinic acid leads to large decrease of the androgenic and the anabolic activity [242]. This may be due to the increased hydrophilic character of the resulting esters which renders the steroids water-soluble, and which in turn may effect the excretion or the attachment by adsorption to the receptor. Hydrophobic steroids may be taken up and held in solution in the body fat. The other factor which has to be taken into account is the action of the steroid on the end organ, which may involve attachment to a protein. This process will determine the biological potency of steroids. However, there is a competition for the steroids by different receptors and it is probable that steroids are taken up by a large number of substances in the body. The degree to which this occurs will be determined by the structure of the steroid. We will deal with the structure of steroids next.

CHAPTER 2

STRUCTURE—ACTIVITY THEORIES

RINGOLD'S THEORY

Ringold was the first investigator to put forward a theory for hormonal activity by systematic examination of the effects created by stereochemical and electronic alterations of the steroid molecule [3]. He postulated that the interaction of androgens with a receptor to produce a classical androgenic response is on the α-face (back side) of the androgen molecule. He reached this conclusion after examining the androgenic action of α- and β-alkyl- or halo-substituted androstanes. Inherent in the conclusion is the fact that the active molecules are essentially planar with the angular methyl substituents attached to the C-10 and C-13 carbon atoms projecting out from the β-face (top side) of the molecule, and the keto group at C-3 and the hydroxyl group at C-17 essentially in the plane of the molecule. The postulation that α-face adsorption is responsible for androgenic activity was supported by the following facts:

(1) Replacing the 17α-hydrogen by a methyl substituent maintains full androgenic activity; however, increasing the length of the 17α-substituent by one carbon unit (the change from methyl to ethyl) results in decrease in androgenic activity, which is attributed to steric factors.

(2) Removal of the 19β-methyl groups (the change from the androstane series to the 19-nor series) does not result in increased androgenic activity. If the β-face of the molecule were responsible for the androgenic activity the absence of the 19β-methyl group could result in an increase in androgenic activity.

(3) Introduction of a 6β-methyl group in the androstane series results in an increase of androgenic activity compared to the unsubstituted androstane derivative, while 6α-substitution does not decrease the activity.

(4) Introduction of a 1α-methyl group in the moderately potent androgen, 19-norandrostan-17β-ol-3-one, eliminates androgenic activity.

(5) The lowered activity of the mono 2α- and 4α-methylandrostanes

cannot be construed as an argument for or against α-face adsorption since the methyl group is in the plane of the carbonyl group of ring A in both cases.

(6) Androstane-$3\alpha,17\beta$-diol is a potent androgen while androstan-$3\beta,17\beta$-diol is a very weak androgen. In this case an axial 3α-hydroxyl substituent would not be a bar to enzymatic interaction since the C-3 oxygen function is probably one of the points of enzymatic attachment.

Despite some reservations outlined by Ringold when he introduced it, the α-face hypothesis with respect to the androgens seemed reasonably convincing. He also commented on the decreased androgenicity of 2,2-dimethyl-, 4,4-dimethyl-, and $2\alpha,6\beta$-dimethylandrostan-17β-ol-3-one. In each case he attributed the decreased potency to severe steric interactions, which cause distortion of ring A. Further, he pointed out that the activity of B-homoandrostan-17β-ol-3-one and 8-isotestosterone, in which rings B and C are cis-annulated, indicates that in the androgen series steric factors in the center of the molecule are not as crucial as distortion in rings A or D.

Ringold also considered the role of substitution by electronegative atoms. He called attention to the fact that in the androgen series, as opposed to the progestational and corticoid hormones, the α,β-unsaturated ketone group, i.e., the Δ^4–3-ketone system, is not required for high hormonal activity and saturated 3-ketoandrostanes (5α-H) are potent agents as well as are the 3α-hydroxyandrostanes. Ringold attributed this finding to the fact that if the C-3 oxygen function is involved in an enzymatic or cellular combination, which he felt is almost certainly the case, the firmness of this binding cannot be of as great importance for the androstanes as for the progestational and corticoid hormones. To support this theory, he pointed out that the substitution of a powerful electronegative group (such as 2α-fluoro or 4-chloro) has a marked influence on the Δ^4–3-ketone system by effecting the polarization of this system. This effect may cause the Δ^4–3-ketone system to resemble more a saturated 3-ketone system. Hence the explanation that while the 4-chloro-substituted progestational agent and corticoids are inactive, the 4-chloro-substituted androgens are active. The role of 2α-fluoro substitution was also explained by polarization factors, but the role of 6-fluoro substitution remained unexplained.

ZAFFARONI'S DATA

Zaffaroni [78] studied the effects of alkyl and electronegative group substitution on the activity of androgenic–anabolic steroids. His findings for androgenic activity confirm Ringold's postulation; in addition he included anabolic activities. The main points of his data may be summarized as follows:

(1) 1α-Methyl-19-nordihydrotestosterone is inactive as an androgen and 20% as active as testosterone as an anabolic steroid.

(2) The 2α-methyl substitution decreases the androgenic activity and increases the anabolic activity of the parent steroid. 2α-Methyldihydrotestosterone was found to be as active orally as 17α-methyltestosterone as an anabolic agent and to have very low androgenicity. $2\alpha,17\alpha$-Dimethyldihydrotestosterone possesses 20% of the androgenic and 400% of the anabolic activity of 17α-methyltestosterone when administered orally.

(3) 3-Methyl substitution (α or β) completely eliminates androgenic and anabolic activity.

(4) 4-Methyltestosterone was found to be 40% as androgenic and 120% as anabolic as testosterone. 4α-Methyldihydrostestosterone was 5–10% as androgenic and 100% as anabolic; 4β-methyldihydrostestosterone, 5–10% as androgenic and 200% as anabolic as testosterone.

(5) 6α-Methyl substitution in the testosterone and dihydrotestosterone series decreases the androgenic potency and increases the anabolic potency while 6β-methyl substitution increases both activities.

(6) 2,2-Dimethyl, 4,4-dimethyl, or $2\alpha,6\beta$-dimethyl substitution eliminates activity.

(7) 2-Hydroxymethylene substitution in the 17α-methyldihydrotestosterone series decreases the androgenic activity to 20% that of 17α-methyltestosterone while increasing the anabolic activity to 400% that of the same compound.

(8) 4-Chlorotestosterone exhibits 50% of the androgenic and 100% of the anabolic activity of testosterone.

(9) 6α-Fluorotestosterone possesses 50% of the androgenic and 100% of the anabolic activity of testosterone, while 6β-fluorotestosterone has 25% of the androgenic and 25% of the anabolic activity of testosterone.

(10) 6α- and 6β-Chloro substitution diminishes both the androgenic and anabolic activities of testosterone.

As can be seen, Zaffaroni's data complement Ringold's theory.

BUSH'S THEORY

Bush [4], however, disagreed with Ringold's theory. According to Bush [4] up to 1962 "the balance of the evidence is in favor of an association of the α-side of the steroid with receptors for androgenic activity, but more work is needed before the uncertainties about the role of substituents at C-17 will be resolved. A β-sided association is suggested, for instance, by the activity of epimeric 17-hydroxysteroids." Bush's main criticism of Ringold's theory, which favors α-sided attachment of androgens to their receptor, centered around the uncertainty of the effects of 17α-substituents. As Bush outlines, "the 17α-methyl group, while conferring oral activity on a typical C_{19}17β-ol, reduces androgenic potency by parenteral routes while leaving levator activity intact." Larger 17α-alkyl groups reduce or abolish androgenic activity, though most clinical workers have found that androgenic activity is still present. Bush also called attention to the fact that in androgens substitutions and modifications in ring A have much more complicated and striking effects than in other classes of steroids. In ring A α-substitutions appear to cause considerable losses of androgenic activity, such as in compounds containing 2α-methyl, 1α-acetylthio, and 3α-methyl groups. On the other hand, 9α-bromosteroids have fair or considerable androgenic potency. Bush also suggested that the considerable activity of 6α-methyl- and 4-methyl-testosterones can be attributed to the fact that close apposition of these steroids to their receptor does not occur over the lower edge of rings A and B. Furthermore the appreciable androgenic activity of 8α-testosterone suggests that close apposition of the β-surface of ring B is not involved in the steroid–receptor association for androgen activity. Bush pointed out the fact that a number of 3-deoxy and 3β-fluoro-3-deoxysteroids possess significant androgenic activity. He suggested that the 17β-hydroxyl group plays a specific role in the association of this group with the receptors for levator and androgenic activity, since all active steroids in this group possess a 17β-hydroxyl group or else a group which can be converted to it (17-ketone or 17α-hydroxyl). According to Bush, all these data confirm the validity of the broader concept outlined by him — that changes in the configuration of the upper half of rings C and D have a far greater effect on all types of biological activity of steroids than many changes in the configuration of rings A and B.

In a later article [7], Bush pointed out that analysis of the structure–activity relationship suggested that in the case of androgens, probably, but not certainly "the upper and β-sided surface of the molecule was responsible for conferring specificity on the steroid–receptor association." According to Bush the importance of the configuration of the "upper" half of the rings (carbons 1,2,11,12,13,16,17) is evidenced by the preparation of potent androgens and anabolic agents possessing a heterocyclic ring [11] attached to ring A. These new compounds have shown that the Δ^4–3-ketone system is not necessary for androgenic activity and a large coplanar extension of the steroid nucleus from the A-ring "leftwards" is compatible with androgenic activity. Bush favored [7] a β-sided attachment of testosterone to its receptors even though 17α-alkyl substituents larger than the ethyl group cause a sharp decrease in androgenic or levator activity. That several 17α-methyl and 17α-ethyl derivatives possess considerable activity points to the fact, according to Bush, that the α-surface of the steroid is not involved in a *specific* interaction with the receptor, since 17α-methyl and 17α-ethyl substitutions already bring about considerable distortion of the molecular surface of the steroid.

Bush [4] did not favor a specific interaction of the steroid hormones with their receptors. The fact that synthetic steroids which differ from the natural hormones display hormonal activity leads him to the conclusion that those regions of the natural steroid hormones which can be modified in the synthetic steroids without loss of activity are not responsible for close association with their receptors, i.e., "these regions are either not associated closely with their receptors or . . . they associate with a flexible part of their receptors in a way that confers little or no specificity upon the interaction."

In Bush's explanation [4] of the nature of drug–receptor interaction the association between steroid and receptor does not involve chemical reaction. He also ruled out the possibility that the high biological activity of certain analogs is due to metabolic modification. He attributed the intrinsic action of the steroids on their receptors or their surroundings to some effect of a physical nature. According to Bush the 17β-hydroxyl group, which appears to be the crucial polar group in the natural androgen testosterone, increases the specificity of receptor association by increasing the probability that the steroid attaches itself by the correct phase. This is similar to Sarett's concept [180] of anti-inflammatory steroid action. Once this one-point, loose attachment to the receptor is made by one of its polar groups, rotation around this point can bring the molecule into correct apposition with the rest of the receptor surface. As soon as the correct apposition of the steroid to the receptor is reached, the association becomes strong enough to permit little or no movement of the parts of the molecules that are in close apposition. Thereby the receptor is prevented by the presence of steroids from interacting with other molecules or groups.

This mechanism was defined later by Bush [6] as the blocking mechanism. Bush's classification is based on Paton's rate theory of drug action [181]. Paton suggested that a pharmacological effect depends not upon the number of individual drug–receptor combinations but upon the rate, k_1, of drug–receptor occupation. He assumes that receptor occupation itself immobilizes receptors, so that the observed pharmacological effect depends upon the ratio k_2/k_1, where k_2 is the rate of dissociation of drug–receptor complexes. When k_2 is high (case 1) the drug will be a powerful stimulant (agonist) because receptors are rapidly set free from occupation; when k_2 is low (case 2) the drug will be mainly antagonistic and its biological effect may be explained by the fact that its occupancy of receptor sites prevents the agonists from access to these specific receptor sites. Intermediate values of k_2 characterize partial agonists.

In Bush's turnover mechanism k_2 is high (Paton's case 1); it requires the association of a receptor with the drug involving chemical reaction or transport, followed by the dissociation of the second molecule (carrier or exchange mechanism) or of a very similar (chemical change) molecule. The rate-limiting step is determined by the rate constant for the dissociation reaction.

On the other hand, in the blocking mechanism k_2 is low (Paton's case 2); the rate-limiting step is determined by the association of steroid with the receptor. The association blocks the receptors against the approach and association of agonists. According to Bush [6], the blocking mechanism must involve either the competitive inhibition of the turnover mechanism or the production of physical changes on the molecular scale. Steroid hormones produce their action merely by sitting down on the receptors, they are in the class of "blocking agents" rather than "turnover agents." This would confirm Bush's earlier postulation [4] that the reactive groups of steroid hormones are not essential as *reactive* groups for steroid hormone activity (since no chemical reaction takes place).

THEORY OF BOWERS AND CO-WORKERS

In 1963 a Syntex group [43] set out to study the structural requirements of androgenic–anabolic activity. Their concept was to vary the electron density pattern in and around ring A. According to Bowers, in conjunction with steric properties the electron density pattern should be related to the ability to bind to proteinlike structures. They found that Δ^1-, Δ^2-, and Δ^3-17β-hydroxy-5α-androstenes all exhibit reasonable anabolic activity, the anabolic–androgenic ratio being more favorable in the latter two compounds. The corresponding Δ^4-olefin exhibited only very weak activity and Δ^3-17β-hydroxy-5β-androstene was inactive. 17α-Methyl-17β-hydroxy-Δ^2-5α-androstene had 50% of the androgenic and 200% of the anabolic activity of 17α-methyltestosterone, when administered orally. According to Bowers, other data also aid the conclusion that a high electron density at C-2 and/or C-3 in 17β-hydroxyandrostane is a factor strongly promoting high myotrophic activity. This condition may be satisfied by a C-3 carbonyl or oxidizable C-3 hydroxyl group. It is possible that C-3 ketones may be active primarily as enols or enolate anions where a Δ^2 π-bond is present. Bowers also postulated that "introduction of more than one sp^2-hybridized carbon atom into ring A results in a pronounced flattening of the ring from a cyclohexane chair form to a more planar conformation in which the steroid may be better able to rest on a receptor surface with a concomitant increase in the degree of orbital overlap."

While Bowers *et al.* center their attention on the structural requirements of ring A, the more difficult question still remains, What are the steric requirements of androgen–receptor complex interactions? The contradiction still exists between Ringold's α-sided and Bush's β-sided attachment of androgens to their receptors.

THEORY OF WOLFF AND CO-WORKERS

Wolff *et al.* in 1964 suggested on the basis of biological evaluation of 11 steroids having cyclopropane, ethylene oxide, or spirooxiranyl rings fused to C-2 and C-3, that a requirement for androgenic activity is the presence of high electron density at C-2 and/or C-3 such as is provided by sp^2 hybridization. This high electron density is responsible for the formation of a π-complex with the receptor site. Wolff suggests that the steroid is in contact with the receptor surface in two discrete areas: the β-face of rings A, B, and C is attached to a receptor area and the α-face of ring D is attached to a second receptor area. The 13β- and 17β-substituents are in a relatively unhindered environment. It was proposed that the two principal binding sites are the A ring, where a π-bond is formed, and the 17β-function, which can be attached by any of several types of nonbonded interactions. The remaining areas in contact with the receptor would form ordinary hydrophobic bonds or van der Waals bonds. It was suggested that no chemical reaction takes place, but that the effect of the steroid is to induce a conformational change in the receptor. In an earlier paper Wolff [99] rationalized the low androgenic and myotrophic potency of C-19 functional steroids (10-cyano, 10-formyl, and 10-hydroxymethyl compounds) in terms of steric interference with drug–receptor fit. He put forward the hypothesis that androgenic and myotrophic responses are initiated by the β-face absorption of a steroid on a tissue receptor.

NEUMANN AND WIECHERT'S DATA

In 1965, Neumann and Wiechert [100] published the results of their investigations of some 150 steroids. They studied the androgenic and anabolic activity of steroids substituted in the C-1 and/or the C-2 position. Five parent compounds were chosen: 1. 5α-androstan-17β-ol; 2. testosterone; 3. 5α-dihydrotestosterone; 4. Δ¹-5α-androsten-17β-ol 3-one; and 5. Δ²-5α-androsten-17β-ol. The parent compounds were further classified according to the substitution in the 17-position, i.e., 17β-hydroxy, 17α-methyl 17β-hydroxy, or 17β-acetoxy compounds.

Their results may be summarized as follows:

(1) The effect of 1α-methyl substitution was indiscriminate — increased, decreased, or unchanged anabolic and androgenic potency was observed.

(2) With two exceptions the 1β-methyl substitution decreases anabolic activity and significantly androgenic activity. Δ²-Androstenes were the exceptions, for which distortion of ring A causes the normally equatorially oriented β-substituent to assume a quasi-axial position.

(3) 1,1-Dimethyl and 1,1-ethylene substitution brings about almost total loss of androgenic and anabolic activity. 1-Ethyl substitution in Δ¹-compounds also leads to inactivation.

(4) 1-Methylene substitution has no effect on the activity of dihydrotestosterone and 17α-methyl-Δ²-5α-androstene-17β-ol, and decreases the activity in other cases.

(5) 1,2α-Methylene substitution increases both the anabolic activity and the androgenic activity of 17β-acetoxy-5α-androstane while producing a favorable ratio. In all other cases 1,2α-methylene substitution decreases the anabolic activity to a small extent and the androgenic activity to a great extent, thereby producing a favorable ratio.

(6) 1-Methyl substitution of Δ¹-5α-androstenes has no effect on the anabolic activity, but decreases significantly the androgenic activity, thereby producing a very favorable ratio.

(7) 1α-Chloromethyl, 1α-bromomethyl, and 1α-iodomethyl substitution bring about a significant decrease in activities. Since they decrease the androgenicity to a greater extent, the ratio becomes favorable.

(8) 1α-Hydroxy substitution increases both activities of dihydrotestosterone and dihydrotestosterone acetate significantly, while it has no effect on Δ²-5α-androstene-17β-ol 17-acetate.

(9) 1-Ketone substitution decreases the activities.

(10) 1α-Cyano substitution leads to inactivation.

Further studies were carried out with disubstituted substances in order to establish the effects of additional substitutions. The results are summarized as follows:

(1) 7α-Methyl substitution in 1α-methyl-substituted compounds brings about a large increase in both anabolic and androgenic activities, while 7β-substitution leads to inactivation.

(2) Removal of the 19-methyl group (changing to the 19-nor series) in 1α-methyl, 1α-hydroxyl, or $1,2\alpha$-methylene compounds decreases the androgenicity to a great extent while decreasing the anabolic activity only to a smaller extent. Thus the ratio becomes favorable.

(3) 11β-Hydroxy substitution in 1-methyl Δ^1-compounds, 1α-methyl or $1,2\alpha$-methylene compounds brings about a decrease in potencies.

(4) 2-Chloro substitution in 1-methyl Δ^1-compounds decreases the androgenic potency significantly while decreasing the anabolic potency only slightly, thereby producing a favorable ratio.

(5) 3-Chloro substitution in Δ^2-compounds, the introduction of Δ^6-unsaturation, 2β-bromo, or 4-hydroxymethylene substitution all lead to inactivation.

EVALUATION OF NEUMANN AND WIECHERT'S DATA

Considering the question of α-face or β-face attachment of ring A to the receptor surface, the balance of supporting data is *in favor of α-face attachment*. While 1α-methyl substitution is indiscriminate, the decrease in activities caused by 1β-methyl substitution would favor β-face attachment. In contrast, the decrease in activities by 1α-chloromethyl, 1α-bromomethyl, 1α-iodomethyl, and 1α-cyano substitutions all favor α-face attachment. The increase of activities caused by 1α-hydroxy substitution also suggests α-face attachment, since the 1α-hydroxyl group is usually regarded because of its polar nature as a point of attachment to the receptor site. In addition, the multi-substituted compounds suggest that attachment to the receptor on the β-face at C-11 is not important since 11β-hydroxy substitution brings about a decrease in potencies. The multi-substituted compounds also suggest that since removal of the bulky C-19 methyl group (transfer to the 19-nor series) does not increase the potencies, attachment to the receptor on the β-face at C-19 is not critically important. If it were, introduction of a C-19 methyl group would decrease the potencies, but in fact it increases them. The importance of the 17β-hydroxyl group is well documented by the fact that acetylation of the 17β-hydroxyl group of $1,17\alpha$-dimethyl-Δ^1-5α-androsten-17β-ol-3-one completely inactivates both potencies. The role of electron clouds may become important in explaining the increased potencies of $1,2\alpha$-methylene- and 1-oxo-substituted 17β-acetoxy-5α-androstanes. It is reasonable to assume that these increases are due to introduction of electron clouds into the molecule.

KLIMSTRA, NUTTING, AND COUNSELL'S THEORY

In 1966, Klimstra, Nutting, and Counsell [87] dealt a serious blow to Wolff's hypothesis. They examined the validity of the following statement of Wolff's hypothesis: "the tendency is for the asymmetrical sp^2 system to afford a more active compound when hindrance is greater on the α-face, than when hindrance is greater on the β-face" [71]. According to Klimstra, Nutting, and Counsell "this hypothesis is set forth to rationalize the increased activity or lack of it of the α over the corresponding β-isomers of steroidal A ring epoxy androgens." In Klimstra and co-workers' experiments "similar isomeric epoxides appear to have the *reverse order of androgenic and myotrophic activity* when administered parenterally." For example, $2\beta,3\beta$-epoxy-5α-androstan-17β-ol and 17α-methyl-$2\beta,3\beta$-epoxy-5α-androstane-17β-ol were several times more potent than the corresponding α-isomers, i.e., $2\alpha,3\alpha$-epoxy-5α-androstan-17β-ol and 17α-methyl-$2\alpha,3\alpha$-epoxy-5α-androstane-17β-ol. According to Klimstra *et al.* "this would tend to indicate that the more active epoxides have greater hindrance on the β- rather than on the α-face of the tetracyclic system." These results are somewhat in line with α-face (back side) attack of the androgen molecule by the enzymatic surface as proposed by Ringold.

Klimstra, Nutting, and Counsell concluded that "the different results reported as well as the variety of hypotheses expounded seem to further point up the lack of understanding of how and where these substances are bound to receptor sites."

This is certainly a dramatic conclusion to the existing hypotheses of the mode of interaction of hormones with receptors and of the problem of structure and mechanism of action of steroid hormones.

EVALUATION OF BIOLOGICAL DATA

To say that considerable difficulty is encountered in comparing literature data expressing anabolic and androgenic properties would be an understatement. The methods of biological evaluation are given in detail by Dorfman [182, 183]. While androgens have a pronounced effect in humans by influencing the growth of the male genital tract and the development of secondary sex characteristics, androgenicity is most often expressed in terms of the following indices: the growth of the seminal vesicles and ventral prostate in rodents and the growth of the chick and capon comb. (A most recent communication by Cavallero [105] recommends the use of the exorbital lacrimal gland of the castrated rat as an index of androgenicity.) The effect of anabolic activity in humans is associated with nitrogen retention and positive protein metabolism. While Stafford and co-workers [191] and Arnold *et al.* [184] make use of the nitrogen-retaining activity of castrated male rats, and even the nitrogen-retaining activity in the monkey has been suggested by Stucki *et al.* [192] as a measure of anabolic activity, generally the anabolic activity has been expressed in terms of increase in weight of the levator ani muscle of the rat, following the demonstration by Eisenberg and Gordan [185] that the levator ani of the rat responded to testosterone by an increase in weight. Such growth is sometimes referred to as levator activity or myotrophic activity. The main points of criticism with regard to the growth of the levator ani muscle as an indicator of the anabolic activity of steroids are summarized by Hayes [186]. Comparative evaluation of the anabolic activity by nitrogen retention or myotrophic activity is provided by Arnold and Potts [44] and by Potts *et al.* [41].

Meaningful experimental evaluation of anabolic and androgenic activity was started by Hershberger *et al.* [187]. The difference in weight between the ventral prostate (seminal vesicles) of the treated and untreated animals gives a measure of the androgenic activity while the difference in weight of the levator ani muscle gives a measure of the anabolic (or myotrophic) activity of the steroids. Overbeek [42] pointed out the fact that sometimes the weight of the levator ani muscle is divided by the weight of the seminal vesicles to express anabolic activity. This practice is inadmissible according to Overbeek, since their quotients will generally be dose-dependent. He also cautioned that in comparing esters, one can only compare those which have the same duration of action.

In this study, where possible, anabolic activity is expressed as the ratio of the activity of the compound to that of testosterone or testosterone propionate (in most cases subcutaneous administration of the drug) or methyltestosterone (in most cases oral administration of the drug) in increasing the weight of the levator ani muscle of the rat. The standard of comparison (T = testosterone, TP = testosterone propionate, MT = methyltestosterone) is expressed as 100% activity. Androgenic activity where possible, is expressed as the ratio of the activity of the compound compared to that of testosterone or testosterone propionate (subcutaneous administration) or methyltestosterone (oral administration) in increasing the weight of the seminal vesicles or ventral prostate of the rat.

The *anabolic/androgenic ratio (quotient), Q,* is expressed as the anabolic activity of the compound compared to T, TP, or MT as measured by the percent increase in weight of the levator ani muscle in the castrated rat divided by the androgenic activity of the compound compared to that of T, TP, or MT as measured by the percent increase in weight of the seminal vesicles or ventral prostate of castrated rat.

CHAPTER 3

A NEW APPROACH TO THE

STRUCTURE—ACTIVITY RELATIONSHIP

In all previous hypotheses for the mechanism of action of hormones, the main consideration has been restricted to the question whether the α- or the β-face of the steroid is involved in the steroid–receptor interaction. These theories are evaluated on the basis of changes in hormonal activities brought about by structural modifications. Since obviously the search is for compounds with high activity, most of the theories are derived from results brought about by optimal structural modifications.

Since we feel that perhaps better results can be obtained from another viewpoint, we propose to consider the mechanism of action by an approach considering the *minimal structural requirements necessary to bring about hormonal activity*. We must remember that the natural androgen and anabolic agent, testosterone, does not display the minimum structural requirements of active steroids — there are many steroids which are less active; testosterone itself is an excellent androgen and a good anabolic agent.

Starting with the basic compound which displays the minimal structural requirements, we will consider the effects of the following alterations on the hormonal activity:

1. Ring size
2. Stereochemistry of annulation, configurational, and conformational changes
3. Replacement of carbon by other elements
4. Replacement of hydrogen by other elements and functional groups and introduction of unsaturation

5α-ANDROSTANE — BASIC STRUCTURE

In 1960, the very fundamental postulation was made by Segaloff and Gabbard [22] that *androgenic activity is a property of the hydrocarbon steroid skeleton* of the natural androgen, 5α-androstane. It might be

33

argued that certain seco steroids possessing only three rings also display androgenic properties, such as Wilds' [195] (E-46), Dorfman's [21] (E-6), or Jacques' [117] (E-13 and E-14) compounds.

Granting that a tricyclic ring system takes precedence over a tetracyclic ring in simplicity, the other structural modifications, namely introduction of oxygen functions and unsaturation, are high order deviations from the simple hydrocarbon. However, we are concerned only with compounds containing the steroid skeleton.

Before Segaloff and Gabbard's results showing that a hydrocarbon steroid containing no oxygen exhibits androgenic property were published in 1960, Huggins and Jensen [188] had found that androstan-17β-ol, a steroid containing only one oxygen, caused growth of the prostatic gland. Steroids possessing an oxygen function at either position 3 or position 17 (but not at both sites) were also investigated by Kochakian [189], who found that none of these steroids had biological activity when administered by implantation as pellets. However, he found that the absorption of these compounds from the subcutaneous tissues was low. The fact that poor solubility was the reason for the inactivity of these compounds was later demonstrated by Kochakian [19]. He found that the completely saturated A-ring deoxy compound, 17-methylandrostan-17β-ol, displayed high activity in promoting growth of the kidney, the prostate, and seminal vesicles, when administered orally. This finding was later confirmed [86], inasmuch as the compound possessed significant oral anabolic (280% of methyltestosterone) and androgenic (50% of methyltestosterone) activity.

Having defined 5α-androstane as our basic structure, we are now in a position to consider the effects of the modifications outlined above.

THE EFFECT OF RING SIZE

An excellent review of ring contraction is given in Djerassi's "Steroid Reactions" [190]. Examination of some biological data suggests that ring contraction significantly decreases or abolishes androgenic–anabolic activity. The complete elucidation of the effect of ring contraction is beyond the scope of our investigations, inasmuch as ring-contracted steroids do not adapt themselves to the principle of minimal structural requirement. We will, however, consider the effect of ring expansion.

A-Homo Steroids

A-Homotestosterone propionate (compound E-20) is reported [152] to have androgenic and myotrophic potency. The activity is less than, but of the same order of magnitude, as that of testosterone propionate; the ratio of myotrophic to androgenic activity is significantly higher in the case of the A-homo compound than with the testosterone propionate. The A-homo ring can assume two possible conformations. In one of these the C-2 and C-2' carbons project below the plane, and the new C-1 — C-2 and C-2' — C-3 bonds become axial with respect to the rest of the skeleton. It can also be seen that in the newly adopted conformation of the C-1 and C-2 pair becomes completely eclipsed with respect to each other along the C-1 — C-2 axis. The same applies to the newly adopted conformation of the C-2' and C-3 pair and finally to the conformation of the C-2 — C-2' pair. On the other hand the A-homo ring can assume a quasi-boat conformation with the C-2 and C-2' carbons projecting above the plane of the molecule. In this case the new C-2 — C-2' bond becomes quasi-equatorial with respect to the rest of the skeleton and the newly adopted conformations of the C-1 and C-2, C-2 and C-2', and C-2' and C-3 pairs along their respective axes are no longer eclipsed. Therefore the second conformation is more favored. These data will yield additional evidence that α-face adsorption is really the case on ring A.

In either case the molecular dimensions of ring A are not greatly affected in one respect. Consider the distance between the two sides of ring A, namely the C-1 — C-2 side and the C-3 — C-4 side. From molecular models it can be seen that insertion of the C-2' atom into the hexagon (making a heptagon) will not greatly alter the distance between the C-3 — C-4 side (this remained unchanged) and the imaginary C-2 — C-3 axis (which corresponds to the old C-2 — C-3 side). In other words the "width" of the A ring will remain the same. As will be seen later, this is an important consideration.

Examining now A-homodihydrotestosterone (compound E-10) it is seen that the conformation of ring A is quite different from that in A-homotestosterone. Carbon atoms C-4 and C-5 are now sp^3-hybridized as opposed to the sp^2 hybridization of those atoms in the first case. Inspection of molecular models will reveal that ring A can adopt several conformations. In one case carbons C-2 and C-2', and in the second case carbons C-2' and C-3, project above the plane, while in two other conformations C-2 and C-2', or C-2' and C-3 project below the plane. However by far the most important conformation A-homodihydrotestosterone can adopt will be the one in which the C-2 and C-3 carbon atoms project below the plane with C-2' also projecting, but to a slighter degree below the plane of the molecule. Although the "width" of ring A in this case, as already outlined before, remains the same, projection of ring A below the plane of the molecule becomes the overriding issue and explains the loss of activity, as will be explained later. This fact points to the importance of α-face absorption at ring A.

B-Homo Steroids

Ringold examined [111] the activity of B-homoandrostan-17β-ol-3-one in which ring B contains seven instead of the six carbon atoms found in normal steroids. This compound is as androgenic as androstan-11β-ol-3-one, while the anabolic activity is increased. Assuming that there is no possibility for the B-homo ring to take up a boat conformation, there are still seven conformations possible for ring B, four different chair forms with planes of symmetry passing through carbons 5, 8, 9, or 10 and three twist forms with axes of symmetry passing through carbons 6, 7, and 7'. The conformation of B-homoandrostan-17β-ol-3-one was depicted [3] as having the C-7 and C-7' carbon atoms projecting below the plane of molecule. If we assume that ring B is in a chair form, then the plane of symmetry in the above conformation would pass through C-10. This conformation, however, was pictured later by Sorm *et al.* [380] as unlikely, because of large angle strains and distortions. Sorm expects three conformations to be the most favorable. In two of these, the plane of symmetry passes through carbons 5 and 8, respectively, and in the third possible conformation ring B assumes a twist chair form with an axis of symmetry passing through C-7. Since in the first two conformations there is considerable protrusion of carbons comprising ring B below the plane of the molecule and the compound still possess considerable activity, α-face attachment of the molecule at ring B cannot be an important factor. However, the molecular dimensions of ring B are not greatly affected in one respect. Consider the distance between two sides of ring B, namely the distance between the C-9—C-10 side and the C-6—C-7 side. Inspection of molecular models reveals that insertion of the C-7' atom into the hexagon will not greatly alter the distance between the C-9—C-10 side (this remained unchanged) and the imaginary C-6—C-7 axis (which corresponds to the old C-6—C-7 side). In other words, the "width" of ring B will remain the same.

D-Homo Steroids

Due to the trans junction of rings C and D in natural steroids, the five-membered D ring is strained. Birch [193] pointed out the fact that the 18-CH$_3$ (between rings C and D) appears to be essential for high activity but this effect may be due merely to the fact that it prevents the (natural) trans junction of rings C and D from assuming the more stable cis configuration. The methyl group can be replaced by ethyl group [77] without diminution in activity. With two fused six-membered rings, the trans junction is the more stable and if ring D of the natural hormones is converted into a six-membered ring, the compounds obtained are active, so it may be possible to omit the 18-CH$_3$ group in such cases [162, 165, 194]. W. S. Johnson also expressed the view that the 18-nor compounds may be expected to exhibit activity provided the C/D rings remain transfused, a condition which, in the case of 17-keto compounds, is more apt to be satisfied with a six-membered than with a five-membered ring D.

THE STEREOCHEMISTRY OF ANNULATION AND CONFIGURATIONAL AND CONFORMATIONAL CHANGES

A/B Ring Junction

Etiocholane derivatives in which the C-5 hydrogen occupies the β-position and rings A and B possess a cis ring junction are inactive as androgens or anabolic agents. In these molecules rings A and B are not planar; ring A is bent so that the C-2, C-3, and C-4 carbon atoms project below the plane of the rest of the molecule.

Since in the past the mechanism of hormone action was always interpreted in terms of α- or β-face attachment it is hard to visualize any other but α-face attachment at ring A. Considering that lack of planarity and protrusion towards the α-face abolishes activity, this fact very strongly supports α-face attachment at ring A.

B/C Ring Junction

8-Isotestosterone

8-Isotestosterone (8α, 9α-testosterone) is a stereoisomer of the natural hormone testosterone [16]. The configuration at C-8 center of 8-isotestosterone (compound E-7) is inverted, since the C-8 hydrogen now assumes an α-orientation in contrast to the natural hormones, where the C-8 hydrogen has a β-orientation. As a result of the configurational change at the C-8 center, the B/C ring junction becomes a cis junction. Inversion of configuration at C-8 therefore causes a major conformational change since either ring B or ring C must now assume a boat conformation, as pointed out by the striking changes found in the rotatory dispersion of 8-iso stereoisomers [196]. Therefore, the 8-iso stereoisomers no

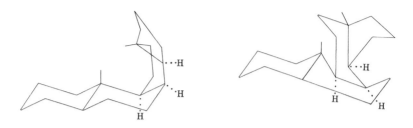

Ring C in boat form   Ring B in boat form

longer possess the natural planar geometry; instead the two planar halves made up from ring A and ring B on one side and ring C and ring D on the other side are perpendicular to each other. This explains the observed decrease in the biological activity of $8\alpha,9\alpha$-testosterone (compound E-7). We would like to point out that the iso steroids may be pictured as possessing either one of the two possible conformations which are probably of similar energy content (with ring B or ring C in the boat form). Assuming that the energy barrier between the two conformations is not too large, both conformations will be significantly populated. The transition in flipping over from one conformation to another conformation must involve a state where the conformation will approach near planarity. On statistical grounds, the population of this planar form must also contribute to the stereochemical picture. If we accept that the planar form exhibits activity, we can justify the observation that isotestosterone possesses some androgenic activity.

Retrosteroids

Stereoisomers possessing the $9\beta,10\alpha$-configuration are called retrosteroids [197]. These compounds have the hydrogen atom on carbon atom 9 in the β-position and the C-19 methyl group on carbon atom 10 in the α-position, i.e., just the reverse of the normal $9\alpha,10\beta$-structure. The B/C ring junction becomes a cis junction and the angular C-18 methyl group on carbon atom 10 has an α-configuration as opposed to the angular C-19 methyl group which occupies a β-configuration on carbon atom 13. Inspection of molecular models reveals that the $9\beta,10\alpha$-stereoisomers may adopt a conformation with all the rings in the chair form.

Inspection of models also reveals that in this conformation on one hand there is a large nonbonded interaction between the C-19 methyl group and the hydrogens on C-12 and C-14. On the other hand, by flipping over ring C into a boat form, this interaction is greatly relieved while the newly imposed interaction between the C-18 methyl group and the hydrogens on C-8 and C-9 is not so severe. The overall geometry of the molecule as a whole is less bent in the second conformation; it approaches planarity to a greater extent. These compounds are not expected to have large androgenic–anabolic activity.

Retrotestosterone has no androgenic activity [198], but the 6-halo derivatives are reported to have favorable anabolic–androgenic ratios in the following order: 6α-fluoro- (greatest), 6β-chloro-, 6β-fluororetro-testosterone [119]. Introduction of additional unsaturation at C-6 and C-1 has a beneficial effect and a "strong anabolic activity with favorable anabolic–androgenic ratio" is reported for the following compounds in decreasing order [120]: 6-chloro-17β-hydroxy-9β,10α-androsta-1,4,6-trien-3-one 17-acetate (greatest), 6-fluoro-17β-hydroxy-9β,10α-androsta-4,6-dien-3-one 17-acetate, 6-chloro analog, 6-unsubstituted analog.

In addition to the classical tests usually applied for the assay of andro-genic and anabolic activities, another method is reported by the same workers for the evaluation of androgenic properties of progestational compounds in the rat by the female fetal masculinization test [201]. It should be pointed out that inversion of configuration at C-10 does not necessarily bring about abolition of androgenic activity, in view of the new theory, as will be seen later.

8α,10α-Steroids

In this class of compounds the configurations at the C-8 and C-10 centers are both inverted. As a result of this configurational change both the A/B and the B/C ring junctions become cis. As expected on the basis of previous considerations, 8α,10α-testosterone does not show any androgenic–anabolic activity [199, 200].

C/D Ring Junction

A number of steroids having the 13α-configuration and simultaneously a C/D cis ring junction have been reported[202] to be inactive.

Ring Extension

As we have already outlined, ever since Kochakian's [19, 189] and Huggins and Jensen's [188] experiments, it has been recognized that the oxygen function at the C-3 position is not necessary to elicit an androgenic response. Nevertheless, up to 1959, all compounds showing a significantly high androgenic or anabolic activity did have an oxygen function at carbon-3. In 1959 and 1960 Clinton [203, 204] and co-workers reported the preparation of steroidal pyrazoles by fusion of a pyrazole ring to the steroid nucleus at positions 2 and 3. By this modification they wanted to alter the relative distance between substituents at the 3-position and the 17-position and change the type and stability of receptor site bonding. The fusion of the pyrazole ring substitutes an "aromatic" nitrogen for oxygen at the 3-position and thus changes the nucleophilic environment in this area. One obvious change was that the C−N bond distance was lengthened to 1.4 Å as compared to the C−O bond distance of 1.2 Å in normal ketones.

As a result of this first report, a large number of compounds containing an isocyclic or a heterocyclic ring fused to the steroid skeleton were prepared and tested for androgenic and anabolic activity. Included are steroidal isoxazoles [205], thiazoles [94], pyrroles [206], pyrimidines [207, 208], oxazines [209], thiazenone [92], quinoline [107], pyridine [122], and furazane [121], all prepared by the fusion of the heterocyclic ring to ring A. Several bis heterocyclic steroids were also prepared by fusion of two heterocyclic rings to the steroid nucleus at ring A and ring D, such as steroidal dipyrimidines, pyrazolopyrimidines, and dipyrazoles [137,210]. The compounds having an isocyclic ring extension include the steroidal fulvenes [107], cyclohexenone [150], cyclohexane [151].

Only fusion of heterocyclic rings to the 2,3-position proved to be a useful extension of the steroidal nucleus to produce enhanced activity or better dissociation of androgenic–anabolic activity. Isocyclic ring extensions and fusion of a heterocyclic ring to other positions, such as 1,2 (D-98) or 16,17 (E-27, E-31), gave products with decreased or abolished activities.

The explanation may be provided partly by the enhanced nucleophilic character of heterocyclic rings to provide a greater electron cloud for binding to the receptor site and partly by stereochemical considerations. The fusion of heterocyclic rings does not interfere with the binding forces, inasmuch as they do not protrude in the directions which are responsible for interactions. They provide an additional flat area and also favorably contribute to the slight conformational change of the steroid nucleus.

We should also mention that simple nitrogen-containing derivatives of keto groups, such as 3-azo (D-150), 3-azine (D-5), and 3-hydrazone (D-15) derivatives, may also *sometimes* favorably affect the activity or the dissociation of activities, while others work in the opposite direction, such as 3-dimethylhydrazone (S-115, S-116) or 17-N-pyrrolidinium salt derivatives (E-15).

REPLACEMENT OF CARBON BY OTHER ELEMENTS

One of the main areas of research in the past decade included the preparation of aza and oxa steroids produced by replacement of carbon atoms in the steroid skeleton by nitrogen and oxygen. Practically all positions were replaced and the only successful modification of the natural isocyclic skeleton was obtained by replacing the carbon-2 atom by oxygen (E-21, E-22, E-29).

In addition to the skeletal modifications accomplished by replacement of one carbon atom, several attempts have been made to improve hormonal activity by replacement of a whole isocyclic ring by a heterocyclic ring. The isocyclic ring A was replaced by a pyrazole [211], isoxazoline [212], pyrimidine [213], or thiopyrimidine [214] ring; however, all these types of compounds were inactive. In view of the fact that fusion of the same heterocyclic rings in addition to the steroid nucleus led to highly active compounds, this finding emphasizes the importance of the steroid skeleton.

REPLACEMENT OF HYDROGEN BY OTHER ELEMENTS AND FUNCTIONAL GROUPS, AND INTRODUCTION OF UNSATURATION

1-SUBSTITUTION

1α-Substitution

Testosterone. Introduction of 1α-methyl (S-150), 1α-ethyl (S-151), 1α-methyl 17-acetate (S-72), 1α-chloromethyl 17-acetate (S-74), 1α-bromomethyl 17-acetate (S-75), 1α-iodomethyl 17-acetate (S-76), 1,1-dimethyl 17-acetate (S-77), 1α,2α-methylene 17-acetate (S-78), 1α-methyl 4-chloro 17-acetate (S-79), 1α-methyl 7α-methyl 17-acetate (S-80), 1α-methyl 7β-methyl 17-acetate (S-81), 1α-methyl 11β-hydroxy 17-acetate (S-82), 1α,2α-methylene 11β-hydroxy 17-acetate (S-83), 1α-methyl 11-keto 17-acetate (S-84), 1α-methyl Δ^6 17-acetate (S-86), 1α-methyl Δ^6 11-keto 17-acetate (S-88), 1α,2α-methylene Δ^6 17-acetate (S-89), 1α-methyl 4-chloro Δ^6 17-acetate (S-90), 1α,2α-methylene 6-chloro Δ^6 17-acetate (S-91), 1α-cyano 17-ketone (S-37), 1α-methyl 17α-methyl (S-92), 1α-methylthio 17α-methyl (S-46), 1α-ethylthio 17α-methyl (S-47), 1α-acetylthio 17α-methyl (S-48), 1α,7α-dithioethyl 17α-methyl (S-59), 1α,7α-dithiopropyl 17α-methyl (S-60), 1α,7α-diacetylthio 17α-methyl (S-61), and 1α,7α,17α-trimethyl (S-94) all decrease androgenic activity. However the anabolic activity of the same compounds is decreased to a much smaller degree compared to the parent compound and in a few cases it is even increased slightly compared to the parent compound (S-72, S-46). The anabolic–androgenic ratio becomes favorable in many instances. The 7α-substituted compounds comprise a special class with respect to anabolic potency: 7α-substitution in addition to 1α-substitution brings about a many-fold increase in the anabolic potency (S-61, S-80, S-94).

5 α-Dihydrotestosterone. 1α-methyl (D-84), * 1α-ethyl (D-170), 1α-cyano (D-88), 1α-hydroxymethyl (D-91), 1α-nitromethyl (D-92), 1,1-dimethyl (D-93), 1,1-ethylene (D-94), spiro-1α-oxiranyl (D-95), 1α,2α-methylene 4-hydroxymethyl (D-97), 1ξ-formyl (D-87), 1ξ-isothiocyano (D-90), 1,2-Δ^2-pyrazolino (D-98), 1-methylene 17α-methyl (D-125), 1α, 2α-methylene 17α-methyl (D-126), 1,1-17α-trimethyl (D-127), 1α-iso-propyl 17-acetate (D-102), 1-methylene 17-acetate (D-103), 1α-chloro-methyl 17-acetate (D-104), 1α-bromomethyl 17-acetate (D-105), 1α-iodomethyl 17-acetate (D-106), 1-keto 17-acetate (D-109), 1,1-dimethyl 17-acetate (D-110), 1,1-ethylene 17-acetate (D-111), 1,1-spirooxiranyl

---

* 1-Methylene (D-86).

(D-115), 1α,2α-methylene 17-acetate (D-112), 1α,2α-methylene 4-hydroxymethylene 17-acetate (D-113), 1α,2α-methylene 11β-hydroxy 17-acetate (D-114), 1ξ,2ξ-methyl 17-acetate (D-116), 1α-hydroxy 2β-bromo 17-acetate (D-117), 1α,2α-methylene 2β-chloro 17-acetate (D-118), 1α,2α-methylene 2β-bromo 17-acetate (D-119), 1α-methyl 2ξ-acetoxy 17-acetate (D-120), 1α,17α-dimethyl (D-123), 1α-methyl 17-acetate (D-100), 1α,17α-dimethyl 2-hydroxymethylene (D-172), 1α,17α-dimethyl[3,2-c]pyrazole (D-177), and 1-methylene 2,2-dimethyl 17-acetate (D-121) substitutions all decrease androgenic activity compared to the parent compound. However the anabolic activity of the same compounds is decreased to a much smaller degree. Therefore the anabolic–androgenic ratio as a result of this type of substitution becomes favorable in many cases. The 7α-methyl substitution again constitutes a special class of compounds; 1α,7α-dimethyl (D-96), and 1α,7α,17α-trimethyl (D-128), substitutions substantially increase the anabolic activity but do not alter the androgenic activity. 1α-Hydroxy (D-89), 1α-hydroxy 17-acetate (D-107), and 1α-acetoxy 17-acetate (D-108) slightly decrease the androgenic but substantially increase the anabolic potency but, if α-face attachment is important at this position, this is not unexpected in view of the fact that this group is considered to be involved at the site of enzymatic attachment.

19-Nortestosterone. 1-Methyl-Δ¹-3-desoxo-5α-dihydro 17-acetate (N-64), 1α-methyl 5α-dihydro (N-48), 1α-methyl 5α-dihydro 17-acetate (N-61), 1α-hydroxy 5α-dihydro 17-acetate (N-62), 1α,2α-methylene 5α-dihydro 17-acetate (N-63), and 1α-methyl-Δ²-5α-dihydro 17-acetate (N-65) substitutions all markedly decrease the androgenic activity while the anabolic activity is decreased in a smaller degree. The anabolic–androgenic ratio becomes favorable.

5α-Androstan-17β-ol. 1α-Methyl (D-129),* 1α,2α-methylene (D-131), 1α-methyl 17-acetate (D-135), 1α,17-dimethyl (D-139), and 1α,2α-methylene 17α-methyl (D-141) substitutions all markedly decrease the androgenic activity, while the anabolic activity is decreased to a lesser extent. This type of substitution has a favorable effect on the anabolic–androgenic ratio. The 1α,2α-methylene 17-acetate (D-137) has the same androgenic activity as the parent compound, while the anabolic potency is increased and 1-keto 17-acetate (D-138), 1α-hydroxy 17α-methyl (D-59) and 1-keto 17α-methyl (D-56) substitutions increase both potencies to produce a favorable anabolic–androgenic ratio. This finding is not unexpected on the basis of reasoning previously offered. 1α-Methyl-3α-hydroxy (D-171) substitution decreases the androgenic activity and

---

* 1α-Formyl (D-14).

slightly increases the anabolic activity compared to 3α-hydroxy (D-31) substitution.

5α-Androstane-17β-ol with Ring A Unsaturation. 1-Methyl Δ^1 3-keto (A-86), 1-ethyl Δ^1 3-keto (A-87), 1-ethynyl Δ^1 3-keto (A-88), 1-methoxy Δ^1 3-keto (A-89), 1-methyl Δ^1 3-keto 17-acetate (A-6), 1-ethyl Δ^1 3-keto 17-acetate (A-96), 1-chloro Δ^1 3-keto 17-acetate (A-97), 1-acetoxymethyl Δ^1 3-keto 17-acetate (A-98), 1-chloromethyl Δ^1 3-keto 17-acetate (A-99), 1-methyl Δ^1 3-keto 17α-methyl (A-104), 1-methyl Δ^1 3-keto 17α-methyl 17β-acetate (A-105), 1-ethyl Δ^1 3-keto 17α-ethyl (A-106), 1-methyl Δ^1 3-keto 17α-vinyl (A-107), 1-methyl 2-chloro Δ^1 3-keto (A-90), 1-methyl 2-bromo Δ^1 3-keto (A-91), 1-methyl 2-methoxy Δ^1 3-keto (A-92), 1-methyl 4-hydroxymethylene Δ^1 3-keto (A-93), 1-methyl 11β-hydroxy Δ^1 3-keto (A-94), 1-methyl 11α-hydroxy Δ^1 3-keto (A-95), 1-methyl 2-methyl Δ^1 3-keto 17-acetate (A-100), 1-methyl 2-chloro Δ^1 3-keto 17-acetate (A-101), 1-methyl 2-methoxy Δ^1 3-keto 17-acetate (A-102), 1-methyl 11β-hydroxy Δ^1 3-keto 17-acetate (A-103), 1-methyl Δ^1 (A-137), 1-methyl Δ^1 17-acetate (A-138), 1-methyl Δ^1 17α-methyl (A-141), 1-keto Δ^2 (A-79), 1-keto Δ^2 17-acetate (A-122), 1-keto Δ^2 17α-methyl (A-67), 1α-methyl Δ^2 (A-108), 1α-methyl Δ^2 17-acetate (A-115), 1α-methyl Δ^2 17α-methyl (A-72), 1α-methyl Δ^2 17α-ethynyl (A-126), 1α-hydroxy Δ^2 17-acetate (A-117), 1α-hydroxy Δ^2 17α-methyl (A-70), 1-methylene Δ^2 (A-110), 1-methylene Δ^2 17α-methyl (A-132), 1-ethylene Δ^2 (A-112), 1-ethylene Δ^2 17-acetate (A-121), 1-ethylene Δ^2 17α-methyl (A-133), 1,1-dimethyl Δ^2 (A-111), 1,1-dimethyl Δ^2 17-acetate (A-120), 1,1-dimethyl Δ^2 17α-methyl (A-134), 1α-chloro Δ^2 17-acetate (A-118), 1α-cyano Δ^2 17-acetate (A-119), 1-methylene Δ^2 3-chloro 17-acetate (A-124), 1-keto Δ^2 3-methyl 17-acetate (A-125), 1α-methyl Δ^4 (A-136), 1α-methyl Δ^4 17-acetate (A-139), and 1α-methyl Δ^4 Δ^6 17-acetate (A-140) substitutions all decrease the androgenic activity to a large degree. However the anabolic activity in almost all cases is decreased only to a smaller degree. Therefore the anabolic–androgenic ratio is favorably effected. In a few cases the anabolic activity is increased, such as 1α-methyl Δ^2 17α-methyl (A-72), 1-keto Δ^2 17α-methyl (A-67), and 1-methyl Δ^1 3-keto 17-acetate (A-6).

1β-Substitution

Testosterone. 1β-Methyl 17-acetate (S-73) and 1β-acetylthio 17-ketone (S-38) substitutions give a marked decrease in the androgenic activity. The anabolic activity of the first compound is slightly decreased and that of the second compound markedly decreased. The anabolic–androgenic ratio of S-73 is favorable.

5 α-Dihydrotestosterone. 1β-Methyl (D-85), 1β-methyl 17-acetate

(D-101), and 1β-methyl 17α-methyl (D-124) substitutions all markedly decrease the androgenic potency, while the anabolic potency is decreased only to a smaller degree. Hence the anabolic–androgenic ratio becomes favorable.

19-Nortestosterone. The effect of 1β-substitution is not known in this area.

5α-Androstan-17β-ol. 1β-Methyl 17-acetate (D-136) substitution offers only a slight decrease in the already small anabolic activity while not affecting the very small androgenic potency. 1β-Methyl 17α-methyl (D-140) substitution decreases both indices.

5α-Androstan-17β-ol with Ring A Unsaturation. 1β-Methyl Δ^2 (A-109), 1β-methyl Δ^2 17-acetate (A-116), 1β-methyl Δ^2 17α-methyl (A-127), 1β-methyl Δ^2 17α-ethyl (A-128), 1β-methyl Δ^2 17α-vinyl (A-129), bis(1β-methyl Δ^2 17α-yl) (A-131), 1β-methyl Δ^2 17α-ethynyl (A-130), 1β-methyl Δ^2 3-chloro (A-113), 1β-methyl Δ^2 3-chloro 17-acetate (A-123), and 1β-methyl Δ^2 3-chloro 17α-methyl (A-135) substitutions all markedly decrease the androgenic activity while the anabolic potency is decreased only to a smaller degree. This brings about a favorable anabolic–androgenic ratio, especially in the case of A-109 and A-127.

1,2-Unsaturation

Testosterone. Δ^1 (S-70), Δ^1 Δ^6 (S-106), Δ^1 2-chloro 17-acetate (S-110), Δ^1 4-chloro 17-acetate (S-114), Δ^1 1-methyl 17-acetate (S-85), Δ^1 1-methyl Δ^6 17-acetate (S-87), Δ^1 17α-methyl (S-15), Δ^1 2,17α-dimethyl (S-68), Δ^1 2-formyl 17α-methyl (S-137), Δ^1 4-chloro 17-acetate (S-114), Δ^1 7α-mercapto 17α-methyl (S-57), Δ^1 7-acetylthio 17α-methyl (S-58), Δ^1 4-chloro 11β-hydroxy 17α-methyl (S-103), Δ^1 9α-chloro 11β-chloro 17α-methyl (S-127), and Δ^1 9α-bromo 11β-fluoro 17α-methyl (S-128) substitutions all significantly decrease the androgenic activity. At the same time the anabolic activity is decreased only to a smaller degree and in some cases it is even increased (S-127), to bring about in all cases a favorable anabolic–androgenic ratio. Δ^1 7α-Methyl 17α-methyl (S-138) and Δ^1 2-methyl Δ^6 17α-methyl (S-69) substitutions both increase very slightly the androgenic and to a larger degree the anabolic potency to bring about a favorable anabolic–androgenic ratio. This can be explained by the protective effects of 7α-methyl, in one case, and of 2-methyl Δ^6-unsaturation, in the other case, against metabolic inactivation and against configurational changes.

The Overall Effect of 1-Substitution

The evidence presented in the preceding chapters overwhelmingly emphasizes the importance of α-*face attachment of androgens at carbon-1* of the steroid skeleton. 1α-Substitution generally decreases androgenic activity. However, the anabolic activity is decreased only to a smaller degree; hence it is possible to obtain a favorable anabolic–androgenic ratio by the variation of 1α-substituents.

The evidence also points to the importance of the 1β-substitution effect. Thus 1β-substitution decreases the androgenic and – to a smaller degree – the anabolic potency, as well producing a favorable anabolic–androgenic ratio. However the activity changes caused by 1β-substitution cannot be construed as an argument for β-face attachment at carbon-1, since the 1β-substituents are in the plane of the molecule. Instead the *importance of 1β-substitution should be taken into account in favor of the third-dimensional attachment* of the steroid molecule to the receptor, as will be postulated later. The importance of *third-dimensional attachment* becomes evident in case of Δ¹-unsaturated compounds, where the substituent attached to carbon-1 is *clearly no longer an α- or a β-substituent* but protrudes in the plane of the molecule. It can be seen that 1-substitution of Δ¹-unsaturated compounds also decreases the androgenic activity and to a smaller degree the anabolic activity, to provide a favorable anabolic–androgenic ratio.

2-SUBSTITUTION

2α-Substitution

Testosterone.   2α-Fluoro (S-33), 2α-cyano (S-109), 2-keto (S-113), 2α-chloro 4-chloro (S-111), 2α-chloro 6α-chloro (S-112), 2-chloro Δ¹ 17-acetate (S-110), 2α-dimethylamino 17-acetate (S-41), 2α-dimethylamino 17-acetate methiodide (S-42), 2-methylene 17α-methyl (S-32), 2α-methyl 17α-methyl (S-108), 2-methyl Δ¹ 17α-methyl (S-68), 1α,2α-methylene 11β-hydroxy 17-acetate (S-83), 1α,2α-methylene Δ⁶ 17-acetate (S-89), 1α,2α-methylene 6-chloro Δ⁶ 17-acetate (S-91), 2-formyl Δ¹ 17α-methyl (S-137), 1α,2α-methylene-17-acetate (S-98), 2α-fluoro 17α-ethynyl (S-34), 2-hydroxymethylene 17α-ethynyl (S-35), and 2-hydroxymethylene 17α-propynyl (S-36) substitutions all markedly decrease the androgenic activity and to a smaller degree decrease the anabolic activity. Thus this type of structural change leads to a favorable anabolic-androgenic ratio.

5α-Dihydrotestosterone.   2α-Methyl (D-49), 2α-methyl Δ⁹⁽¹¹⁾ (D-50), 2α-hydroxymethyl (D-8), 2-methoxymethylene (D-20), 2-hydroxymethylene (D-147), 2α-fluoro (D-18), 1,2-Δ²-pyrazolino (D-98), 2α-methoxymethyl (D-17), 1α,2α-methylene 17α-methyl (D-126), 2-hydroxy-

methylene 17α-methyl   (D-10),   2-benzoyloxymethylene 17α-methyl
(D-16), 2,2-17α-trimethyl (D-21), 2-N-methylanilinomethylene 17α-
methyl (D-22), 2α,17α-dimethyl $\Delta^{9(11)}$ (D-55), 2-[2'-(N,N-diethylamino)-
ethylaminomethylene] 17α-methyl (D-80), 2-N-piperidylmethylene 17α-
methyl (D-81), 2-N,N-dimethylaminomethylene 17α-methyl (D-82), 2-
N,N-diethylaminomethylene 17α-methyl   (D-83),   2-methylene 17α-
methyl (D-152), 2α-methyl 17α-methyl (D-45), 2α-acetylthio 17α-methyl
(D-44),   2-aminomethylene 17α-methyl   (D-78),   2-[2'-(N,N-dimethyl-
amino)ethylaminomethylene] 17α-methyl (D-79), 2α-formyl 17α-methyl
(D-151), 2α-methyl 17α-methyl 17β-propionate (D-6), 2α,6α,17α-tri-
methyl 17β-propionate (D-7), 1α,2α-methylene 17-acetate (D-112),
1α,2α-methylene 4-hydroxymethylene 17-acetate (D-113), 1α,2α-meth-
ylene 11β-hydroxy 17-acetate (D-114), 1ξ,2ξ-methyl 17-acetate (D-116),
1α,2α-methylene 2β-chloro 17-acetate   (D-118),   1α,2α-methylene 2β-
bromo 17-acetate (D-119), 1α-methyl 2ξ-acetoxy 17-acetate (D-120),
1-methylene 2,2-dimethyl 17-acetate (D-121), 2α-methyl 17-propionate
(D-28), 2α-methyl 17β-[6-hydroxymethyltetrahydropyranyloxy] (D-34),
2α-methyl 17β-tetrahydropyranyloxy (D-35), 2α-methyl 17-methoxy
(D-19), and 2α-methyl 17-dichloroacetoxy (D-158) substitutions all decrease androgenic activity to a great extent. However the anabolic
activity is decreased only to a smaller degree, and is even increased in
some instances, such as D-8, D-10, D-55, D-45, D-6, D-7, D-35, D-158,
D-50, to produce a favorable anabolic–androgenic ratio.

19-Nortestosterone. 2,2,17α-Trimethyl 5α-dihydro (N-23), 3β-
hydroxy 2,2-dimethyl 17α-methyl 5α-dihydro (N-24), Δ^2 5α-dihydro
17α-methyl (N-29), 5α-dihydro 17α-methyl[2,3-d]isoxazole (N-72),
17α-methyl[2,3-d]isoxazole (N-76), 2α-methyl 5α-dihydro (N-22), 2-
hydroxymethyl Δ^2 5α-dihydro (N-31), 2-formyl Δ^2 5α-dihydro (N-32), 1α,
2α-methylene 5α-dihydro 17-acetate (N-63), 1α-methyl Δ^2 5α-dihydro
17-acetate (N-65), 2-cyano Δ^2 5α-dihydro 17-acetate (N-30), and 17-
acetate[2,3-d]isoxazole (N-70) substitutions all decrease androgenic
activity to a large extent while the anabolic activity is decreased only to
a smaller degree; thus the whole effect is a favorable anabolic–androgenic
ratio.

5α-Androstan-17β-ol. 1α,2α-Methylene (D-131), 2α,3α-epoxy (D-
64), 2α-methyl 3β-hydroxy (D-133), 2α-methyl 3α-hydroxy (D-164),
2-methylene (D-157), 1α,2α-methylene 17-acetate (D-137), 2-methylene
3β-acetoxy 17-acetate (D-75), 1α,2α-methylene 17α-methyl (D-141),
2-keto 17α-methyl (D-57), 2-methylene 3β-hydroxy 17α-methyl (D-76),
2-methylene 3β-acetoxy 17α-methyl   (D-77),   2-methylene 17α-methyl
(D-9), 2α,3α-epithio 17α-methyl (D-70), 17α-methyl [2,3-c]oxadiazol

(D-71), 17α-methyl[3,2-c]pyrazole (D-11), 2α,17α-dimethyl 3-azine (D-5), 17α-methyl[3,2-c]isoxazole (D-33), 2α,3α-epoxy 17α-methyl (D-38), 2α,3α-methylene 17α-methyl (D-41), 2α,3α-difluoromethylene 17α-methyl (D-52), 2α-methyl 3-hydrazone 17α-methyl (D-15), 17α-methyl 2,3-cyclohexeno (D-72), 17α-methyl 2,3-cyclohexano (D-73), 17α-methyl[3,2-b]pyridine (D-149), and spiro-2β-oxiranyl (D-42) substitutions all decrease the androgenic activity while in most cases lowering, but only to a smaller degree the anabolic activity. In a few cases (D-9, D-70, D-71, D-5, D-11, D-52, D-33, D-15, D-150), the anabolic activity is increased. All these types of substitution lead to a favorable anabolic–androgenic ratio. In the case of 5α-androstan-17β-ol, which does not possess the 17α-methyl substitution and therefore is not protected against metabolic inactivation at the only site offered for a stronger type of attachment to the receptor, introduction of another "active site" into the molecule may become important enough inasmuch as it may override the normal steroid effect. Therefore, it is not surprising to find that with introduction of an "electron-rich" substituent into this bare molecule, for example by 2α,3α-epoxy 17-acetate (D-37), 2α,3α-methylene (D-40), 2α,3α-epithio (D-69), and 2α,3α-difluoromethylene 17-acetate (D-51) substitutions, the androgenic as well as the anabolic activities are both increased compared to the parent compounds (D-134 and D-145, respectively), with the latter having an edge and producing a favorable anabolic–androgenic ratio.

5α-Androstan-17β-ol with Ring A Unsaturation. 2-Methyl Δ^1 3-keto (A-10), 2-bromo Δ^1 3-keto (A-143), 2-methyl Δ^1 3-keto 17-acetate (A-61), 2-chloro Δ^1 3-keto 17-acetate (A-142), 2-methyl Δ^1 3-keto 17α-methyl (A-11), 2-methyl Δ^1 3-keto 17α-ethyl (A-12), 2-cyano Δ^1 3-keto 17α-methyl (A-42), 2-methylthio Δ^1 3-keto 17α-methyl (A-60), 2-formyl Δ^1 3-keto 17α-methyl (A-63), 2-bromo Δ^1 3-keto 17α-methyl (A-74), 2-chloro Δ^1 3-keto 17α-methyl (A-75), 2-hydroxy Δ^1 3-keto 17α-methyl (A-76), 2-methoxy Δ^1 3-keto 17α-methyl (A-77), 1-methyl 2-chloro Δ^1 3-keto (A-90), 1-methyl 2-bromo Δ^1 3-keto (A-91), 1-methyl 2-methoxy Δ^1 3-keto (A-92), 1-methyl 2-methyl Δ^1 3-keto 17-acetate (A-100), 1-methyl 2-chloro Δ^1 3-keto 17-acetate (A-101), 1-methyl 2-methoxy Δ^1 3-keto 17-acetate (A-102), 2-chloro 3β-hydroxy Δ^1 17α-methyl (A-144), 2-chloro 3β-acetoxy Δ^1 17α-methyl (A-145), 2-methyl Δ^1 17-acetate (A-35), 2-methyl Δ^2 17α-methyl 17β-acetate (A-31), 2-methyl Δ^2 (A-23), 2-methyl Δ^2 17α-methyl (A-24), 2-cyano Δ^2 17-caproate (A-27), 2-cyano Δ^2 (A-44), 2-cyano Δ^2 17α-methyl (A-43), 2-cyano Δ^2 17α-methyl 17β-acetate (A-53), 2-formyl Δ^2 (A-36), 2-formyl Δ^2 17α-methyl (A-28), 2-formyl Δ^2 Δ^4 17-acetate (A-37), 2-difluoromethyl Δ^2 (A-29), 2-difluoromethyl Δ^2 17-

acetate (A-40), 2-fluoromethyl Δ^2 (A-64), 2-fluoromethyl Δ^2 17-acetate (A-41), 2-hydroxymethyl Δ^2 (A-38), 2-hydroxymethyl Δ^2 17α-methyl (A-26), 2-chloromethyl Δ^2 17-acetate (A-39), 2β-hydroxyethyl Δ^2 (A-50), 2-carboxylic Δ^2 17α-methyl (A-62), Δ^2[2,3-d]isoxazole (A-156), $\Delta^2\Delta^4$ [2,3-d]isoxazole (A-149), 17α-methyl $\Delta^2\Delta^4$[2,3-d]isoxazole (A-58), 17-acetoxy $\Delta^2\Delta^4$[2,3-d]isoxazole (A-147), 17α-methyl $\Delta^2\Delta^4\Delta^6$[2,3-d]isoxa-zole (A-148), 17α-methyl Δ^2[2,3-d]isoxazole (A-30), 17α-methyl Δ^2 [3,2-d]thiazole (A-158), 17α-methyl Δ^2[3,2-d]methylthiazole (A-85), 17α-methyl Δ^2[3,2-d]methylpyrimidine (A-84), Δ^2[3,2-e]thiazenone (A-83), Δ^2[3,2-b]quinoline (A-33), 2-acetoxymethylene Δ^3 (A-47), 17-acetoxy 2-keto Δ^3 (A-80), 17α-methyl 2-keto Δ^3 (A-66), 2α-methyl 17α-methyl $\Delta^3\Delta^5$ (A-51), and 17α-methyl Δ^4[3,2-c]pyrazole (A-157) substitu-tions all decrease the androgenic activity to a medium or large degree. In this type of substitution, the anabolic activity is often decreased to a small degree, thereby producing a favorable anabolic–androgenic ratio. But in many cases the anabolic activity is increased. The following are the compounds showing the most favorable anabolic–androgenic ratios: among the Δ^1-unsaturated compounds, A-6, A-10, A-11, A-61, A-63, A-104; and among the Δ^2-unsaturated compounds, A-24, A-27, A-28, A-38, A-43, A-44.

2β-Substitution*

Testosterone. No examples were found for this kind of substitution.

5α-Dihydrotestosterone. 2β-Fluoro (D-153) substitution abolishes both androgenic and anabolic activities.

19-Nortestosterone. No examples were found for this kind of substitution.

5α-Androstan-17β-ol. 2β,3β-Epoxy (D-67), 2β,3β-epithio (D-65), 2β, 3β-epithio 17α-methyl (D-66), and 2β-hydroxy 17α-methyl (D-60) sub-stitutions substantially decrease both the androgenic and anabolic potencies, leaving a favorable anabolic–androgenic ratio. In contrast, however, 2β,3β-epoxy 17-acetate (D-36) and 2β,3β-difluoromethylene 17-acetate (D-53) substitutions substantially increase both the andro-genic and anabolic properties. Moreover, as already mentioned, Klimstra, Nutting and Counsell [87] reported results with respect to the 2β,3β-epoxy 17α-methyl (D-68) and 2α,3α-epoxy 17α-methyl (D-38) pair as well as the 2β,3β-epoxy (D-67) and 2α,3α-epoxy (D-64) pair. Their re-sults indicate that the more active epoxides have greater hindrance on the

---

*2-Disubstituted compounds are listed under 2α-substitution.

β-face rather than on the α-face of the tetracyclic system, in line with α-face attachment at this area.

The Overall Effect of 2-Substitution

In line with the principle of minimal structural requirement, there is no evidence to support a β-face hypothesis at carbon-2. However, the importance of 2α-substitution is well documented. Since, as in the case of 1β-substitution, the 2α-substituents are in the plane of the molecule, the importance of 2α-substitution is a significant support in favor of the third-dimensional attachment of the steroid molecule to the receptor, as will be postulated later. The importance of third-dimensional attachment becomes evident in the case of Δ^1- or Δ^2-unsaturated compounds, where the substituent attached to carbon-2 is no longer an α- or a β-substituent but protrudes again in the plane of the molecule. It can be seen that planar, 2-substitution of Δ^1- or Δ^2-unsaturated compounds also decreases the androgenic activity and to a smaller extent the anabolic activity to provide a favorable anabolic–androgenic ratio. Furthermore, certain substituents in the 2-position not only decrease the androgenic activity but increase the anabolic activity, so judicious selection of substituents can lead to highly favorable dissociation of anabolic and androgenic activity.

3-SUBSTITUTION

3α-Substitution

5α-Androstan-17β-ol. 5α-Androstan-3α,17β-diol (D-31), and 5α-androstan-3α-ol-17-one (D-27), derivatives are very potent androgens, while the anabolic activity is decreased compared to the parent compound, to give an unfavorable anabolic–androgenic ratio. In these cases, an axial 3α-hydroxyl substituent will not interfere with the attachment to the receptor since the C-3 oxygen function is probably one of the points of attachment. 2α,3α-Epoxy (D-64), spiro-3α-oxiranyl (D-142), 3α-hydroxy 3β-methyl (D-163), 3β-hydroxy 3α-methyl (D-162), 2α,3α-epithio 17α-methyl (D-70), 2α,3α-epoxy 17α-methyl (D-38), 2α,3α-methylene 17α-methyl (D-41), and 2α,3α-difluoromethyl 17α-methyl (D-52) substitutions all substantially decrease the androgenic activity. The anabolic activities of D-38, D-41, D-64, and D-142 are small. Ringold [3] accounted for the inactivity of D-162 and D-163. The androgenic activities of the 2α-methyl 3β-hydroxy (D-133) and 2α-methyl 3α-hydroxy (D-145) compounds are not known, but the anabolic property of the former is the same as that of the parent compound while the anabolic potency of the latter is slightly higher than that of the parent compound. The anabolic activities of D-52 and D-70 are greater than that of the parent compound, giving a favorable anabolic–androgenic ratio.

As already outlined under 2α-substitution, $2\alpha,3\alpha$-epoxy 17-acetate (D-37), $2\alpha,3\alpha$-methylene (D-40), $2\alpha,3\alpha$-epithio (D-69), $2\alpha,3\alpha$-difluoromethylene 17-acetate (D-51), and, in addition, $3\alpha,4\alpha$-epoxy 17-acetate (D-39) substitutions all give increased androgenic as well as anabolic activity, with a favorable anabolic–androgenic ratio.

5α-Androstan-17β-ol with Ring A Unsaturation. Both the 3α-hydroxy Δ^4 17-acetate (A-55) and 3α-hydroxy Δ^4 17-(1′-ethoxy)-cyclohexyl ether (A-154) compounds are very potent androgens with high anabolic activity and show an unfavorable anabolic–androgenic ratio.

3β-Substitution

19-Nortestosterone. The reduced form, 3β-hydroxy derivative (N-34) shows decreased androgenic and anabolic potencies with a favorable anabolic–androgenic ratio. The same applies for N-107. The N-26 and N-27 derivatives show a very favorable anabolic–androgenic ratio.

5α-Androstan-17β-ol. We have already dealt with the effects of $2\beta,3\beta$-disubstitutions, such as D-36, D-38, D-53, D-60, D-64, D-65, D-66, D-67, and D-68, under 2-substitution.

3β-Hydroxy (D-146), 3β-hydroxy 17α-methyl (D-61), spiro-3β-oxiranyl (D-43), 3β-cyclopentyl (D-156), and 3β-hydroxy 17-ketone (D-161) substitutions all cause decreased androgenic activity while the anabolic activity is either very small or not recorded at all.

5α-Androstan-17β-ol with Ring A Unsaturation. 3β-Hydroxy Δ^1 (A-13), 3β-hydroxy Δ^1 17α-methyl (A-15), 3β-hydroxy Δ^1 17α-ethyl (A-17), 3β-hydroxy 2-chloro Δ^1 17α-methyl (A-144), 3β-acetoxy 17-acetate (A-14), 3β-acetoxy 17α-methyl (A-16), 3β-acetoxy 17α-ethyl (A-18), and 3β-acetoxy 2-chloro Δ^1 17α-methyl (A-145) substitutions all cause decreased androgenic activity. At the same time, the anabolic activity is decreased to a smaller extent, to yield a favorable anabolic–androgenic ratio.

On the contrary, 3β-hydroxy Δ^4 (A-54) and 3β-hydroxy Δ^4 17α-methyl (A-146) substitutions substantially increase the androgenic potency, while the former also increases to a smaller degree the anabolic potency, thus creating an unfavorable anabolic–androgenic ratio. The effect of 3β-hydroxy Δ^4 17α-methyl substitution (A-146) on the anabolic activity is not known. 3β-Acetoxy Δ^4 17-acetate (A-56) substitution has no effect on the androgenic activity while decreasing the anabolic activity of A-4.

We have already discussed the effects of 2,3-disubstitutions in the Δ^2-unsaturated series under 2α-substitution.

Androst-3,5-dien-17β-ol. 3-Methyl 17α-methyl (A-32), 3-chloro 17-acetate (A-155), 3-cyclohexylenol ether 17α-methyl (A-160), 3-cyclo-pentyl- (A-161) and 3-cyclopentylenol ether 17α-methyl (A-162) sub-stitutions all significantly increase the androgenic activity. The anabolic activity is increased to a smaller degree to produce an unfavorable anabolic–androgenic ratio.

Androst-5-en-17β-ol. 3β-Acetoxy 17α-propargyl(A-165), 3β-hydroxy (A-152), 3β-hydroxy 17α-methyl (A-5), and 3β-hydroxy 17-ketone (A-57) substitutions all bring about a large decrease in androgenic and anabolic potencies, while the anabolic–androgenic ratio remains unfavorable.

Other 3-Substitutions

Replacement of the 3-ketone, as well as the 17β-hydroxyl function, by fluorine, i.e., 3,3,17,17-tetrafluoro-5α-androstane (E-26), abolishes both androgenic and anabolic activities. Simultaneous replacement of the 3-ketone and 17β-hydroxyl group by a heterocyclic ring, i.e., 5α-androstano[3,2-*d*]pyrimidine[17,16-*c*]pyrazole (E-27) and 5α-estrano-[3,2-*c*]pyrazole [17,16-*d*]pyrimidine (E-31), also abolishes both activities.

Testosterone 3-dimethylhydrazone (S-115), 17α-methyltestosterone 3-dimethylhydrazone (S-116), 3-methylene-17α-methylandrost-4-en-17β-ol (S-26), 3-methylene-17α-ethylandrost-4-en-17β-ol (S-27), 9α-fluoro-3-methylene-17α-methylandrost-4-en-11β,17β-diol (S-28), 5α-dihydro-testosterone-3-isonicotinylhydrazone (D-154), 3-methylene-17α-methyl-17β-hydroxyestr-4-ene (N-19), 3-methylene-17α-ethyl-17β-hydroxyestr-4-ene (N-20), and 19-nortestosterone 3-dimethylhydrazone (N-75) all have the androgenic property decreased to a large extent. At the same time, the anabolic potency is also decreased. The anabolic–androgenic ratio becomes favorable only in one case (S-26). 3,3-Azo-17α-methyl-5α-androstan-17β-ol (D-150) shows decreased androgenic and increased ana-bolic activity to produce a very favorable anabolic–androgenic ratio, while 2α,17α-dimethyl-5α-androstan-17β-ol-3-hydrazone (D-15) and 2α,17α-dimethyl-4,5α-dihydrotestosterone-3-azine (D-5) show no change in androgenic activity but increased anabolic activity to yield a favorable anabolic–androgenic ratio. 3-Methylene-17α-methyl-17β-hydroxy-5α-androstane (D-54) has both potencies decreased to the same degree. In 17β-hydroxy-5α-androstane-3-fulvene (D-155), 3β-cyclopentyl-17β-hydroxy-5α-androstane (D-156), 17α-methyl-17β-hydroxy-2β(*H*)-2,3-cyclohex-2'-eno-5α-androstan-4'-one (D-72) and 17α-methyl-17β-hy-

droxy-2β(*H*)-2,3-cyclohexano-5α-androstan-4'-one (D-73) both potencies have been abolished.

The Overall Effect of 3-Substitution

We have already described 5α-androstane, our basic structure displaying androgenic properties; an oxygen function at the carbon-3 position is
not necessary for androgenic or anabolic activity. This postulation is
supported by the androgenic and anabolic activity of a large number of
3-desoxy compounds. Introduction of a 3-oxygen, especially a 3-ketone
function, however, usually enhances activity. Substitution of the 3-ketone
function by methylene or nitrogen derivatives is permitted. Fusion of a
heterocyclic ring to the 2- and 3-positions with simultaneous incorpora-
tion of unsaturation at the new junction leads to a highly successful new
class of compounds exhibiting favorable anabolic–androgenic ratios.

The importance of bond angles, distances, and directions in trying to
assess relations between the structure and physiological activity was
greatly emphasized by Woodward [215]. He pointed out the fact that introduction of the 3-carbonyl function, i.e., the change of cyclohexane
ring A to a 3-cyclohexanone ring A brings about very little distortion from
cyclohexane geometry as far as the backbone structure of the carbons and
the attached groups is concerned. The only change brought about by
introduction of the 3-carbonyl function is the considerable change in the
position of the carbonyl carbon with respect to the rest of the molecule
[216]. The distance from the α-carbons to the carbonyl carbon was found
to be close to 1.50 Å [217, 218] as compared to a normal 1.545 Å carbon–
carbon (single bond) distance, and the carbonyl angle to 116°, as compared
to the normal 109° tetrahedral angle.

Since, stereochemically, not much change is observed by introducing a
carbonyl group into the 3-position of the steroid skeleton, the nature of
the observed enhanced androgenic and anabolic activity is probably due
to electronic factors. Introduction of an electron-rich substituent must
provide additional force to the attachment of steroid to receptor.

Substitution of a methylene group (which has a van der Waals radius of
2.0 Å) for the carbon-3 ketone function changes the bond distance to
1.34 Å [217] from 1.22 Å [218], and the bond angle C—C=C to 122°
from 116° (C—C=O), with the additional distance of the C—H bond
length, 1.08 Å, and the bond angle H—C—H 108° 44' [218]. Similarly,
substitution of nitrogen for the carbon-3 ketone function changes the
bond distance to 1.4 Å [219]. It can be seen that substitution of the
oxygen function by a methylene group or a nitrogen derivative lengthens
the bond distances but, more important, alters the bond angles consider-

ably. Disubstitution in the 2- and 3-positions with the simultaneous introduction of Δ^2-unsaturation will alter the bond lengths and angles even further. This is particularly apparent when a five-membered hetero-cyclic ring is fused to carbon atoms 2 and 3. This change may bring about a very favorable change in the anabolic–androgenic ratio.

The high androgenic activity of 3α-hydroxy-substituted compounds is used as an argument in favor of α-face absorption since the carbon-3 oxygen would probably be one of the points of attachment to the receptor.

In line with the principle of minimal structural requirements, the indiscriminate results of 3β-substitution point to the uninvolvement of the 3β-position. Furthermore, as we have already seen, comparison of $2\beta,3\beta$- and $2\alpha,3\alpha$-disubstituted compounds [87] revealed that the more active epoxides have greater hindrance on the β-face than on the α-face of the tetracyclic system, in line with α-face attachment at the carbon-3 position.

4-SUBSTITUTION

5α-Dihydrotestosterone. 4α-Methyl (D-47), 4β-methyl (D-46), $1\alpha,2\alpha$-methylene 4-hydroxymethylene (D-97), and $1\alpha,2\alpha$-methylene 4-hydroxy-methylene 17-acetate (D-113) substitutions all decrease the androgenic activity to a great extent and the anabolic activity to a smaller degree.

19-Nortestosterone. 4-Hydroxy 17-cyclopentylpropionate (N-77), 4-chloro 17α-methyl (N-12), 4-chloro 17α-methyl 17β-propionate (N-13), 4-methyl (N-15), 4-chloro 17-acetate (N-14), and 4-hydroxy 17-acetate (N-16) substitutions all decrease the androgenic activity. The anabolic activity is only slightly decreased or even increased (N-77), thus the anabolic–androgenic ratio becomes favorable. 4-Hydroxy 17α-methyl (N-74) substitution greatly increases both activities to bring about a favorable ratio. Dimethyl substitutions, e.g., 4,4-dimethyl 3-keto Δ^5 17α-methyl (N-42), and 4,4-dimethyl 3-keto Δ^5 (N-41), abolish both activities. The reason for the effect of 4,4-dimethyl substitution may be explained according to Ringold [3] by the fact that the 4α-methyl group may possess a pseudo-axial configuration due to severe nonbonded interactions on the β-face (particularly 2-H–6-CH$_3$, 6-H–2-CH$_3$, 19-H–6-CH$_3$ interactions).

5α-Androstan-17β-ol. The $3\alpha,4\alpha$-epoxy 17-acetate (D-39) exhibits in-creased androgenic and anabolic potencies with a favorable anabolic–androgenic ratio. This effect again may be due to the introduction of an electron-rich substituent into ring A of a relatively simple molecule, providing another active site for attachment to the receptor. 4β-Hydroxy 17α-methyl (D-62) and 4-keto 17α-methyl (D-58) substitutions decrease both activities to produce a favorable anabolic–androgenic ratio.

5α-Androstan-17β-ol with Ring A Unsaturation. 1-Methyl 4-hydroxy-methylene Δ^1 3-keto (A-93), 4-keto Δ^2 17-acetate (A-81), and 4-ketoΔ^2 17α-methyl (A-68) substitutions all decrease the androgenic and anabolic activities but the last two produce a very favorable anabolic–androgenic ratio. Dimethyl substitutions in 4-positions again abolish activity, i.e., the 4,4-dimethyl 3-ketoΔ^5 (A-34), 4,4-dimethyl 3-keto Δ^5 17α-methyl (A-65), 4,4-dimethyl 3-keto Δ^5 6-methyl 16α-methyl (A-82), and 4,4-dimethyl 3β-hydroxy Δ^5 6-methyl 16α-methyl (A-69) compounds, probably for the same reasons as outlined before.

Testosterone. Δ^1 4-Chloro 17-acetate (S-114), 4-bromo (S-8), 4-methyl (S-11), 4-chloro 11β-hydroxy 17-acetate (S-7), 4-fluoro 17-acetate (S-9), 4-hydroxy 17-acetate (S-10), 1α-methyl 4-chloro 17-acetate (S-79), 4-chloro 17-propionate (S-6), 4-hydroxy 17α-methyl (S-16), 4-chloro 17α-methyl (S-17), 4-mercapto 17α-methyl (S-49), 4-methylthio 17α-methyl (S-50), 4-ethylthio 17α-methyl (S-51), 4-acetylthio 17α-methyl (S-52), and 4-chloro 11β-hydroxy 17α-methyl (S-102) substitutions all decrease the androgenic property. However, the anabolic potency is only slightly decreased and in a few cases increased, such as S-114, S-11, S-16, S-6. Substitution at the 4-position by bulky substituents i.e., 4-ethyl (S-13) and 4-allyl (S-14), abolishes both activities.

The Overall Effect of 4-Substitution

Introduction of the 4,5-double bond into the androstane molecule shortens the 4–5 bond length to 1.34 Å from 1.545 Å. At the same time, the double bond causes the ring A to assume a half-chair conformation. Since the double bond is at a ring junction, five carbon atoms, C-3, C-4, C-5, C-6, and C-10 lie in a plane, C-2 is above the plane, and C-1 is below the plane. Only the bonds extending to substituents at C-1 and C-2 have the character of normal axial or equatorial bonds for cyclohexane but those at C-10 and C-6 are only approximately axial or equatorial with respect to ring A. Energy considerations [220] reveal that the half-chair cyclohexene is slightly less stable than the chair cyclohexane. Electronic considerations become important in the case of additional carbonyl substitution at C-3. The length of the carbonyl group of a molecule was correlated with its first ionization potential; a decrease of the ionization potential means an increase of the negative charge on the oxygen atom, and this in turn means an increase of bond length, probably because of increased single bond character according to the resonance description of the carbonyl bond.

$$\text{C=O} \longleftrightarrow \text{C}^+\text{—O}^-$$

Introduction of unsaturation between carbons 4 and 5 causes inter-action of the donor–acceptor type involving the π-electrons of the α,β-unsaturated carbonyl system. Introduction of α,β-unsaturation to the carbonyl group decreases the ionization potential of the carbonyl group or, in other words, increases the polarity of the carbonyl group. In terms of the resonance theory, the net change is an enlarged contribution of ionic structures, e.g.,

$$O{=}C{-}C{=}C \longleftrightarrow \bar{O}{-}C{=}C{-}C^{+}$$

We may consider now the effect of substitution at C-4 of testos-terone ($O{=}C{-}C{=}C$ system). If there is a hydrogen atom at C-4, the large electronegativity of the oxygen atom gives an acceptor character to the carbonyl group and a donor character to the unsaturated bond. Therefore, the resonance contributing structures depicted below are important:

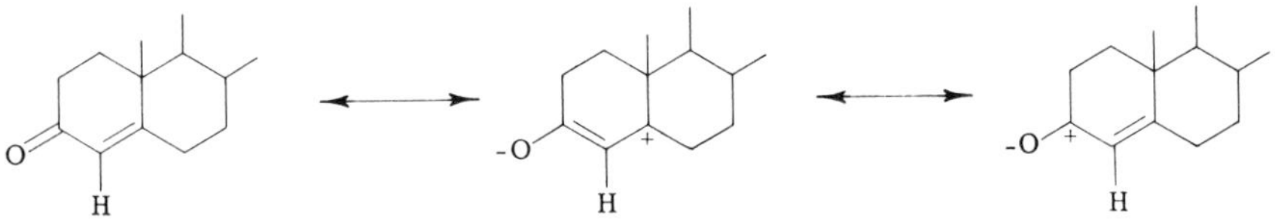

However, if the hydrogen at C-4 is substituted by atoms having lone pairs of electrons (4-chloro, 4-fluoro, 4-hydroxy), resonance contributions such as

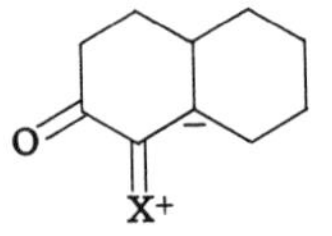

become important. Even 4-methyl substitution can account for such structures by hyperconjugation. Hetero substitution at C-4 therefore de-creases the polarity of the carbonyl group and the stability of the enone system by resonance-contributing structures of the type shown above, which decrease the double bond character of the unsaturated bond.

The most important function of the 4,5-double bond seems to be the protection of the 3-carbonyl function against metabolic inactivation. It is known that 5α-dihydrotestosterone is a more powerful androgen than testosterone, especially when measured by the ventral prostate index.

However, the 3-carbonyl function is metabolized much faster in 5α-dihydrotestosterone than in testosterone where the 4,5-double bond has a rate-limiting effect, i.e., the metabolic inactivation of 3-carbonyl group is much faster than the reduction of 4,5-double bond [168, 221].

Stereochemical factors are only of secondary importance at the carbon-4 position.

(1) Dialkyl substitution in the 5α-dihydro series or Δ^5-series causes a considerable distortion and forces the α-substituent to orient itself in a quasi-axial conformation. Since we have already seen the importance of axial attachment at C-3 and will see the same at C-5, it is not surprising to find that the quasi-axial conformation at C-4 inteferes with the α-face attachment to the receptor at carbons 3 and 5 and hence abolishes activity.

(2) Bulky (ethyl, allyl) substituents in the testosterone series also significantly decrease the activity by causing protrusion exceeding the ordinary distances of simple substituents in the plane of the molecule.

(3) Unsaturation between C-4 and C-5 causes ring A to assume a half-chair conformation. The C-19 methyl group is therefore no longer axial with respect to ring A and will cause a small conformational distortion of the C-19 methyl group with respect to the steroid skeleton as a whole as compared to that in 5α-dihydrotestosterone. This factor will contribute to a *small degree* to the fact that replacement of C-19 methyl group by a C-19 hydrogen (change to the 19-nor series) very often will favorably affect the androgenic and anabolic activities. This effect will be discussed in detail under 10-substitution.

5-SUBSTITUTION

As discussed previously, etiocholane derivatives in which the rings A and B possess a cis junction are inactive as androgens or anabolic agents. Since this type of substitution may be considered an axial, α-face substitution of the 5α-hydrogen by an alkyl group with respect to the ring B, the importance of α-face attachment at carbon-5 becomes evident. However, the 8-isotestosterone and retrotestosterone series do not belong among the 5α-substituted compounds. The importance of α-face attachment is further evidenced by the decreased potency of 5α-substituted derivatives.

5α-Dihydrotestosterone. 5ξ-Methyl (D-26), 5α-cyano (D-24), 5α-carbamido (D-25), and 4α,5α-epoxy 11β-hydroxy 17α-methyl (D-148) substitutions all decrease both activities to a large extent and 5α-methyl

17-propionate substitution (D-143) abolishes androgenic and anabolic activities.

5α-Androstan-17β-ol. 5α-Hydroxy Δ^2 17α-methyl (A-73) substitution decreases both androgenic and anabolic potencies but the resulting compound still shows some potency.

5,6-Unsaturation

The 3-substituted $\Delta^{5,6}$-derivatives have already been considered under 3-substitution. Introduction of 5,6-unsaturation into the saturated 5α-androstan-17β-ol causes decreases in both potencies; the assistance provided by the "electron cloud" of this substitution at this position must not be important. 17β-Hydroxyandrost-5-ene (A-49) and 17α-methyl-17β-hydroxyandrost-5-ene (A-59) both show decreased androgenic and anabolic potencies and an unfavorable ratio compared to the parent saturated compound. Additional unsaturation between carbons 3 and 4 will further decrease the activities. 17β-Hydroxyandrost-3,5-diene (A-150) and 17α-methyl-17β-hydroxyandrost-3,5-diene (A-48) show decreased activities compared to the Δ^5-analogs, A-49 and A-59, or to 17β-hydroxyandrost-3-ene (A-3).

The fact that the unsaturation between carbons 4 and 5 or carbons 5 and 6 imposes the same kind of conformational distortion in compounds possessing the 3-carbonyl group is documented in the 19-nor series by the comparison of activities of corresponding $\Delta^{4,5}$- and $\Delta^{5,6}$-compounds. The 19-nor 3-keto, Δ^4 17α-methyl 17β-ol (N-4) and 19-nor 3-keto $\Delta^{5,6}$ 17α-methyl 17β-ol (N-39); 19-nor 3-keto Δ^4 17α-ethyl 17β-ol (N-5) and 19-nor 3-keto $\Delta^{5,6}$ 17α-ethyl 17β-ol (N-38); 19-nor 3-keto Δ^4 17α-ethynyl 17β-ol (N-7) and 19-nor 3-keto $\Delta^{5,6}$ 17α-ethynyl 17β-ol (N-37) pairs exhibit the same androgenic and anabolic activities, respectively.

5,10-Unsaturation

The 17α-methyl-17β-hydroxy-$\Delta^{5(10)}$-estren-3-one (N-40) and 17α-methyl-17β-hydroxy-Δ^4-estren-3-one (N-4); 17α-ethyl-17β-hydroxy-$\Delta^{5(10)}$-estren-3-one (N-93) and 17α-ethyl-17β-hydroxy-Δ^4-estren-3-one (N-5); and 17α-ethynyl-17β-hydroxy-$\Delta^{5(10)}$-estren-3-one (N-43) and 17α-ethynyl-17β-hydroxy-Δ^4-estren-3-one (N-7) pairs of compounds were compared and it was found that the $\Delta^{5(10)}$ analogs possess substantially lower androgenic and anabolic activities than the corresponding Δ^4 analogs. Introduction of 5,10-unsaturation brings carbons 1, 4, 5, 6, 9, and 10 into one plane with carbons 3 and 7 projecting below and carbon-8 projecting above the plane. C-2 is virtually in the plane. This type of conformational distortion has an unfavorable effect on the biological activities.

6-SUBSTITUTION

6α-Substitution

Testosterone. 6α-Methyl (S-12), 6α-nitro (S-40), 6-chloro Δ^6 17-acetate (S-64), Δ^6 17α-methyl (S-107), and 6α-methyl 17α-methyl (S-96) substitutions all decrease the androgenic and anabolic activities. Several 1α-substituted Δ^6-testosterone derivatives are considered under 1α-substitution. All Δ^6-unsaturated derivatives reduce the androgenic and anabolic activities compared to the 6,7-saturated analogs (see also Wettstein [2301]. The anabolic–androgenic ratio is favorable in S-64 and S-12. There are conflicting reports for 6α-chlorotestosterone acetate (S-29); Kincl and Dorfman [56] report a larger, and Cross *et al.* [53] report smaller androgenic potency than that of testosterone, while the anabolic activity is greater in both cases. The resulting anabolic–androgenic ratio is favorable. There are more conflicting reports on 6α-fluorotestosterone (S-31). Zaffaroni [78] reported the androgenic activity of 6α-fluorotestosterone to be 50% of that of testosterone and the anabolic activity 100% that of testosterone, while Kincl and Dorfman first reported [56] 200% (ventral prostate), 170% (seminal vesicle), 190% (levator ani) values, later Kincl and Dorfman [82] reported 123% (ventral prostate), 128% (seminal vesicles), and 206% (levator ani) values compared to testosterone (100% on each index). We should point out that the steric requirements of substituents as measured by the van der Waals radii are the following [222]: hydrogen, 1.2 Å; fluorine, 1.35 Å; chlorine, 1.9 Å. As we can see, the fluorine is not much larger and it is possible that the steric effects exhibited by the fluorine as compared to the hydrogen are too small to consider.

5α-Dihydrotestosterone. 6α-Methyldihydrotestosterone (D-4) is reported [30] to have decreased androgenic and anabolic activities with a favorable anabolic–androgenic ratio (disregarding a preliminary report [34] from 1957). 6α-Methyl 17α-methyl (D-13) and 2α,6α,17α-trimethyl 17β-propionate (D-7) substituted compounds both show decreased androgenic activity, with the former showing a slightly decreased anabolic activity and the latter an increased anabolic activity. The anabolic–androgenic ratio is favorable.

$\Delta^{5(6)}$-*Androsten-17β-ol.* 4,4,6,16α-Tetramethyl 3-keto (A-82) and 4,4,6,16α-tetramethyl 3β-hydroxy (A-69) substituted compounds both have very small androgenic activities; the anabolic activities are not known.

19-Nortestosterone. 6α-Methylnortestosterone (N-88) shows decreased androgenic and anabolic activities, with a slightly favorable ratio.

6β-Substitution

Testosterone. 6β-Fluoro (S-62), 6β-methyl (S-95), 6β-nitro 17-acetate (S-39), 6β-chloro 17-acetate (S-71), and 6β-methyl 17α-methyl (S-97) substitutions all bring about a large decrease in the androgenic activity and in the anabolic activity. The ratio is not favorable. The 6β-acetoxy 17-acetate (S-117) has a greatly reduced anabolic activity while the androgenic potency is not known.

5α-Dihydrotestosterone. 6β-Methyl (D-3) substitution causes a decrease in both androgenic and anabolic potencies. Preliminary data [34] were later corrected (compare Kincl [107]).

5α-Androstan-17β-ol with Ring A Unsaturation. 6β-Methyl Δ^1 3-keto 17α-methyl (A-19), 6β-methyl Δ^1 3-keto 17α-ethyl (A-20), and 6β-methyl Δ^2 17α-methyl (A-71) substitutions all cause a great decrease in androgenic and anabolic potencies to produce a favorable anabolic–androgenic ratio.

The Effect of 6,7-Unsaturation

Comparison of 17α-methyltestosterone (S-2), 17α-methyl-5α-dihydrotestosterone (D-2), and 17α-methyl-Δ^6-testosterone (S-107) revealed that the presence of the double bond in the 4,5-position enhanced the anabolic activity and did not alter the androgenic activity, according to Beyler *et al.* [124]. Additional unsaturation lowered both anabolic and androgenic activities, but decreased the anabolic activity to a lesser degree.

Comparison of 17α-methyl-17β-hydroxy-5α-androstano[3,2-c]pyrazole (D-11), 17α-methyl-17β-hydroxyandrost-4-eno[3,2-c]pyrazole (A-157), and 17α-methyl-17β-hydroxyandrost-4,6-dieno[3,2-c]pyrazole (A-168) showed that as the degree of unsaturation is increased there is a corresponding decrease in the anabolic and androgenic activities, with the last compound (A-168) having complete loss of these activities. It was suggested [124] that the anabolic and androgenic properties of this compound were lost due to the flattening imposed by the steroidal diene system. It was also found that introduction of an additional double bond between carbons 6 and 7 in 4-chlorotestosterone acetate (S-148) results in the complete loss of activity (compare with S-146).

The Overall Effect of 6-Substitution

Since substitution of the β-face, axial hydrogen by other substituents brings about a decrease in potency, the importance of β-face attachment to the receptor at C-6 is emphasized. We also notice the decreased androgenic potency of 6α-substituted derivatives. Since the 6α-hydrogen is in the plane of the molecule, the importance of third-dimensional attachment becomes evident. This is particularly evident in Δ^6-unsaturated compounds where the 6-substituent is no longer α- or β-oriented but clearly projects in the plane of the molecule.

7-SUBSTITUTION

7α-Substitution

Testosterone. 7α-Methyl (S-21), 7α-methylthio 17-acetate (S-119), 7α-mercapto 17-acetate (S-121), and 1α,7α-dimethyl 17-acetate (S-80) compared to 1α-methyl 17-acetate (S-72); 7α-methyl 17-ketone (S-23) compared to 17-ketone (S-22); 7α-methyl Δ^1 17α-methyl (S-138), 7α-mercapto Δ^1 17α-methyl (S-57), and 7α-acetylthio Δ^1 17α-methyl (S-58) compared to Δ^1 17α-methyl (S-15); and 7α-methyl 17α-methyl (S-20) substitutions all increase both potencies, giving rise very often to a favorable anabolic–androgenic ratio. 7α-Acetylthio 17-acetate (S-120), 7α-mercapto 17α-methyl (S-53), 7α-methylthio 17α-methyl (S-54), 7α-ethylthio 17α-methyl (S-55), and 7α-acetylthio 17α-methyl (S-56) substitutions cause decrease of both androgenic and anabolic potencies while maintaining a favorable anabolic–androgenic ratio.

5α-Dihydrotestosterone. 1α,7α-Dimethyl (D-96), 1α,7α,17α-trimethyl (D-128), and 1α,7α-dimethyl 17-acetate (D-122) substitutions all cause increase in both activities with a resulting favorable anabolic–androgenic ratio.

19-Nortestosterone. 7α,17α-Dimethyl (N-10), 7α-methyl 17α-ethynyl (N-73), 7α-methyl 17-ketone (N-18), 7α-methyl (N-11), 7α-methyl 17-acetate (N-9), 13β-ethyl 7α-methyl (N-52), 13β,17α-diethyl 7α-methyl (N-53), 13β-ethyl 7α-methyl 17α-ethynyl (N-54), and $\Delta^9\Delta^{11}$ 7α,17α-dimethyl (N-102) substitutions bring about a great increase in the anabolic and androgenic properties, producing a favorable anabolic–androgenic ratio.

7β-Substitution

Only two compounds have been studied containing a 7β-methyl group: 1α-methyl-7β-methyltestosterone acetate (S-81) and 1α,7β-dimethyl-5α-

dihydrotestosterone acetate (D-32). Both decrease the androgenic and the anabolic activities to a very large degree. This finding is especially significant in the view of the large increase obtained in activities by 7α-methyl substitution of the same compounds.

The Overall Effect of 7-Substitution

In view of the large increases in the androgenic and anabolic potencies of 7α-methyl-substituted compounds, α-face attachment to the receptor at carbon-7 is not involved. On the contrary, the increase in activity was attributed to protection of other structural elements (4,5 double bond, 19-norhydrogen) against metabolic inactivation. However, 7β-substitution is important. Since this substituent is in the plane of the molecule, third-dimensional attachment, as in the case of 6α-substitution, is emphasized.

8β-SUBSTITUTION

8-Isotestosterone and 8α,10α-testosterone can be considered as 8β-substituted testosterone derivatives where the 8-H is substituted by an alkyl group. In the case of 8α,10α-testosterone other structural features are prohibitive for biological activity. However, 8-isotestosterone displays decreased, but definite androgenic and anabolic properties. Nagata [160] prepared some 8β-substituted testosterone homologs — 8-methyltestosterone (E-30) and others. In all cases decreased androgenic and anabolic potencies were observed. These facts point to the importance of β-face attachment to the receptor at carbon-8.

9-SUBSTITUTION

Retrotestosterone (9β,10α-testosterone) has the hydrogen atom on carbon atom 9 in the β-position and ring B has replaced the hydrogen normally found in the α-position. As we have seen, some substituted retrosteroids still possess androgenic and anabolic activity.

Testosterone. $\Delta^1$9α-Bromo 11β-chloro (S-123), 9α-chloro 11β-fluoro (S-125), Δ^1 9α-bromo 11β-fluoro 17-propionate (S-124), Δ^1 9α-chloro 11β-chloro 17α-methyl (S-127), Δ^1 9α-bromo 11β-fluoro 17α-methyl (S-128), 9α-bromo 11β-fluoro 17α-methyl (S-126), 3-methylene 9α-fluoro 11β-hydroxy 17α-methyl (S-28), 9α-fluoro 11β-hydroxy 16α-methyl 17α-methyl (S-132), and $\Delta^{9(11)}$ 16α-methyl 17α-methyl (S-131) substitutions all lead to decreased androgenic potencies and, with the exception of S-127, to decreased anabolic potencies. The anabolic–androgenic ratio is often favorable. The androgenic potency of the 9α-fluoro 11β-hydroxy 17-

acetate (S-122) is not known but possesses small anabolic activity. However, 9α-fluoro 11β-hydroxy 17α-methyl (S-19) and 9α-fluoro 11β-oxo (S-25) substitutions both increase the androgenic and anabolic activities to a great degree. Both compounds possess a very favorable anabolic–androgenic ratio. We should point out at this time that the steric requirements of the fluorine atom [222] as measured by the van der Waals radii are not very much different from those of the hydrogen atom (1.35 Å versus 1.2 Å).

The Effect of 9,10-Unsaturation ($\Delta^{9,10}$-Nortestosterone)

$\Delta^{9,10}$-Nortestosterone (N-95), 17α-methyl-$\Delta^{9,10}$-nortestosterone (N-96) subcutaneously, and 17α-ethyl-$\Delta^{9,10}$-nortestosterone (N-97) all exhibit greatly decreased androgenic activities while the anabolic activities are increased. 17α-Methyl-$\Delta^{9,10}$-nortestosterone (N-96) orally shows increased androgenic and anabolic activities. The anabolic–androgenic ratio is very favorable in all cases.

This heteroannular dienone system causes rings A and B to assume a half-chair conformation. Carbon atoms 1, 3, 4, 5, 6, 8, 9, 10, 11 are all in one plane while carbon atom 2 projects above and carbon atom 7 below the plane. The rest of the molecule is below the plane. Addition of the second unsaturation to the Δ^4–3-ketone enone system provides an additional electron-rich unsaturated residue, i.e., it enhances the electron delocalization. The result will be an enlarged contribution of ionic resonance structures such as

$$O=C-C=C-C=C \overset{-}{\longleftrightarrow} O-C=C-C=C-\overset{+}{C}$$

Stereochemically, while the hydrogens attached to C-2 and C-7 still constitute an axial and equatorial bond, the bond which attaches the hydrogen atom to C-8 is only approximately axial with respect to ring B. The nonbonded interaction exerted by C-9 and C-10 substituents is removed. These conformational changes have a favorable effect on the anabolic–androgenic ratio.

The Effect of 9,10- and 11,12-Unsaturation
($\Delta^{9(10)}$, $\Delta^{11(12)}$-Nortestosterone)

Total synthetic studies at Roussel-UCLAF produced a large number of compounds having this type of extended unsaturation. Δ^9,Δ^{11}-Nortestosterone 17-acetate (N-103), Δ^9,Δ^{11}-nortestosterone 17β-methoxymethyl ether (N-104), Δ^9,Δ^{11}-nortestosterone 17β-decanoate (N-105), Δ^9,Δ^{11}-nortestosterone 17-carbobenzoxy-ate (N-106), Δ^9,Δ^{11}-17α-methyl-

nortestosterone   (N-99),   and Δ^9,Δ^{11}-$7\alpha,17\alpha$-dimethylnortestosterone
(N-102) derivatives have both androgenic and anabolic activities in-
creased to a very large extent. The anabolic–androgenic ratio is 1. Electronic considerations reveal that the increased stability of the heteroannular trienone system is caused by enhanced electron delocalization.
Stereochemical considerations reveal that carbons, 1, 3, 4, 5, 6, 8, 9, 10,
11, 12, and 13 are in one plane. Rings A, B, and C assume a half-chair
conformation. Only C-2 is above the plane, still possessing an axial and
equatorial hydrogen bond, while carbons 7, 14, 15, 16, and 17 are below
the plane. C-8 no longer has a true axial bond to a hydrogen atom with
respect to rings B and C. In addition, the C-10 to 19-methyl bond is no
longer truely axial with respect to ring C. The effect of all this conformational change is a large increase in the androgenic and anabolic activity.

   Without disputing the magnitude of change in biological activity caused
by conformational distortion, we should also point out that removal of the
two equatorial and axial hydrogens at C-11 and C-12 and introduction of
two new hydrogens which lie in the plane of the extended unsaturated
system contributes to a large activity change. While the contribution of
this factor to the overall change may be small in magnitude, this fact
points to the importance of substituents at carbons 11 and 12.

The Overall Effect of 9-Substitution

   The facts that inversion of configuration at C-9 is possible without
abolishing the activity and that removal of the hydrogen (introduction of
unsaturation between carbons 9 and 10) sometimes decreases (N-96)
the androgenic and anabolic potencies and in many cases increases both
potencies, point to the lack of importance of attachment to the receptor
at carbon-9. Substitution of the 9-hydrogen by bulkier substituents
usually decreases the activity but in the case of fluorine it increases it
(S-19 and S-25). Removal of the 9-hydrogen and the simultaneous introduction of unsaturation between carbons 11 and 12 greatly enhances
the activities. These facts also point to the uninvolvement of C-9 in the
attachment to the receptor site.

10-SUBSTITUTION

   The retrosteroids ($9\beta,10\alpha$-testosterone derivatives) can be pictured as
compounds having an exchanged configuration at C-10. The C-19 methyl
group which is normally a β-substituent now becomes an α-substituent,
and ring A takes the place of the attachment previously held by the
methyl group. The change in steric environment is not too drastic; at
C-10 the methyl group has been exchanged by an alkyl substituent. Dis-

regarding other stereochemical effects, the change in configuration at C-10 alone will not destroy activity.

On the other hand, in the case of $8\alpha,10\alpha$-testosterone not only is the configuration changed at C-10 but also at C-8. Both A/B and B/C ring junctions become cis and, as can be seen from models, ring A with respect to ring B can now be pictured as a very bulky substituent protruding in the β-face.

All these facts point to the importance of β-face attachment at C-10.

Testosterone–Nortestosterone

It is known [222] that the methyl group has a van der Waals radius of 2 Å while the hydrogen atom has a van der Waals radius of 1.2 Å. Replacement of the 19-methyl group by a hydrogen does not necessarily involve an increase in the androgenic and/or anabolic activity. For example, 1-methyl-17β-acetoxy-5α-androst-1-en-3-one (A-6) has a small but definite androgenic activity and a large anabolic activity and exhibits a very favorable anabolic–androgenic ratio. Substitution of the 19-methyl group by hydrogen abolishes both activities, i.e., 1-methyl-17β-acetoxy-5α-estr-1-en-3-one (N-64) does not display any androgenic or anabolic activity.

Similarly 1α-methyl-17β-acetoxy-5α-androstan-3-one (D-100), 1α-hydroxy-17β-acetoxy-5α-androstan-3-one (D-107), and $1\alpha,2\alpha$-methylene-17β-acetoxy-5α-androstan-3-one (D-112) exhibit substantial androgenic and anabolic activities with very favorable anabolic–androgenic ratios. Still, replacement of the 19-methyl group by hydrogen will bring about a large decrease in the androgenic and anabolic activities, i.e., for the compounds 1α-methyl-17β-acetoxy-5α-estran-3-one (N-61), 1α-hydroxy-17β-acetoxy-5α-estran-3-one (N-62), and $1\alpha,2\alpha$-methylene-17β-acetoxy-5α-estran-3-one (N-63).

On the other hand, 17α-methyl-$4,17\beta$-dihydroxyandrost-4-en-3-one (S-16) possesses 50% of the androgenic and 150–300% of the anabolic activity of 17α-methyltestosterone and yet removal of the 19-methyl group brings about a very large increase in both potencies, i.e., 17α-methyl-$4,17\beta$-dihydroxyestr-4-en-3-one (N-74) possesses 300–1000% of the androgenic and 1300% of the anabolic potency of 17α-methyltestosterone. Similarly, 19-nortestosterone 17-acetate (N-3) has higher androgenic and anabolic activities than testosterone 17-acetate (S-66). In addition, the 7α-methyl group increases the androgenic and anabolic activities of nortestosterone derivatives to a much larger extent than those of testosterone derivatives; the resulting 7α-methylnortestosterone derivatives will have greater androgenic and anabolic activities than those of 7α-methyltestosterone derivatives. Compare N-10 to S-20, N-11 to S-21, N-18 to S-23.

Nortestosterone. 10-Hydroxymethyl Δ² 4,5α-dihydro 17α-methyl (N-56), 10-hydroxy Δ² 4,5α-dihydro 17α-methyl (N-49), 10-hydroxymethyl Δ² 4,5α-dihydro (N-55), 10-cyano 2,3β-epoxy 4,5α-dihydro 17-acetate (N-44), 5α-fluoro 10-hydroxy 4,5α-dihydro 17α-ethynyl (N-69), 10-cyano 17-ketone (N-58), 10-hydroxy (N-68), 10-ethyl (N-101), 10-formyl 3β-hydroxy 4,5α-dihydro Δ⁵ (N-59), 10-hydroxymethyl 3β-hydroxy 4,5α-dihydro Δ⁵ 17α-methyl (N-60), and 10-cyano 3β-hydroxy 4,5α-dihydro Δ⁵ (N-57) substitutions all lead to compounds possessing very small androgenic and anabolic activities. 10-Vinyl (N-100) substitution produces an activity intermediate between testosterone and 10-ethyl-testosterone in both indices. The steric requirement of the vinyl group is between those of the methyl and ethyl groups.

Overall Effect of 10-Substutition

Bulky substituents in 10-positions decrease the androgenic and anabolic activities to a large extent. This definitely points to the β-face attachment at C-10. Results obtained with 9α,10α-testosterone derivatives confirm the importance of β-face attachment. However the steric requirement of the methyl group, as measured by the 2.0 Å van der Waals radius [222], is no bar to β-face attachment to the receptor at C-10. The steric requirements at C-10 can accommodate 10-methyl substitution (the presence of the 19-methyl group).

11-SUBSTITUTION

11α-Substitution

11α-Methyl-11β-hydroxytestosterone (S-139), 11α,17α-dimethyl-11β, 17β-dihydroxy-5α-androstan-3-one (D-159), 11α,17α-dimethyl-3β,11β, 17β-trihydroxy-5α-androstane (D-160), and 1-methyl-11α,17β-dihydroxy-5α-androst-1-en-3-one (A-95) compounds show no androgenic and anabolic activity.

11-Keto Substitution

1α-Methyl-Δ⁶-11-ketotestosterone acetate (S-88), 1α-methyl-11-keto-testosterone acetate (S-84), 3,11,17-triketo-Δ⁴-androstene (adrenosterone) (S-43), and 11-keto-17α-methyltestosterone (S-24) all show decreased androgenic and anabolic potencies compared to the those 11-unsubstituted compound. However, introduction of a 9α-fluoro substituent into S-24 will increase both androgenic and anabolic activities to a large degree to produce a favorable anabolic–androgenic ratio (9α-fluoro-11-keto-17α-methyltestosterone, S-25).

11β-Substitution

Testosterone. Δ¹ 9α-Bromo 11β-chloro (S-123), 9α-chloro 11β-fluoro (S-125), 11β-hydroxy (S-5), 1α-methyl 11β-hydroxy 17-acetate (S-82), 1α,2α-methylene 11β-hydroxy 17-acetate (S-83), 4-chloro 11β-hydroxy 17-acetate (S-7), Δ¹ 9α-bromo 11β-fluoro 17-propionate (S-124), 11β-hydroxy 17-ketone (S-44), Δ¹ 4-chloro 11β-hydroxy 17α-methyl (S-103), Δ¹ 9α-bromo 11β-fluoro 17α-methyl (S-128), and 3-methylene 9α-fluoro 11β-hydroxy 17α-methyl (S-28) substitutions all decrease both androgenic and anabolic potencies. In some cases, the anabolic–androgenic ratio is favorable (S-124, S-83, S-7, S-127). However, in 9α-fluoro-11β-hydroxy-17α-methyltestosterone (S-19) the androgenic and anabolic potencies have been greatly increased over that of 17α-methyltestosterone, while the anabolic–androgenic ratio has become 2. 11β-Hydroxy-17α-methyltestosterone (S-18) and Δ¹ 9α-chloro 11β-chloro 17α-methyl (S-127) show a slightly decreased androgenic potency compared to that of 17α-methyltestosterone, while the anabolic potency has been increased by a factor of 3.

5α-Dihydrotestosterone.   1α,2α-Methylene 11β-hydroxy 17-acetate (D-114) substitution decreases the androgenic and anabolic potencies compared to 1α,2α-methylene 17-acetate (D-112) substitution.

19-Nortestosterone. The 11β-hydroxy 17α-methyl (N-66) derivative compared to 17α-methylnortestosterone (N-4) and the 11β-hydroxy 17α-ethyl (N-67) derivative compared to 17α-ethylnortestosterone (N-5) show increased androgenic and anabolic activities and, in the first case, a favorable anabolic–androgenic ratio.

The Overall Effect of 11-Substitution

11α-Substitutions decrease both androgenic and anabolic potencies. Since the 11α-substituent is in the plane of the molecule this points out the importance of third-dimensional attachment to the receptor at C-11.

The effect of 11β-substitution can be divided in two categories. In the presence of the 19-methyl substituent the primary role of the 11β-substituent is steric hindrance. Bulky substituents (including 11β-hydroxy groups) impose very severe steric interference with 19-methyl and 18-methyl groups and therefore interfere with the β-face attachment at the 19-methyl and 18-methyl positions. In the absence of the 19-methyl substituent, this interference is not present. Therefore, 11β-hydroxy substitution, by virtue of its attachment on the β-face to the receptor, increases the androgenic and anabolic activities in the 19-nortestosterone series.

Introduction of a 9α-fluoro substituent (S-19) into 17α-methyl-11β-hydroxytestosterone (S-18) will increase the androgenic and anabolic activities. The fact that this increase is due to the 9α-fluoro substituent becomes evident by comparing the effect of 9α-fluoro substitution (S-25) into 17α-methyl-11-ketotestosterone (S-24).

12-SUBSTITUTION

This effect is not known since no 12-substituted compounds are known. In view of the new hypothesis, however, we predict that β-substitution at C_{12} will decrease both androgenic and anabolic activities.

13-SUBSTITUTION

13β-Ethyl-17β-hydroxygon-4-en-3-one (N-45) compared to 19-nortestosterone (N-1); 17α,13β-diethyl-17β-hydroxygon-4-en-3-one (N-46) compared to 17α-ethyl-19-nortestosterone (N-5); 17α,13β-diethyl-17β-hydroxygon-4-ene (N-47) compared to 17α-ethyl-17β-hydroxyestr-4-ene (N-8); 17α-ethynyl-13β-ethyl-17β-hydroxygon-4-en-3-one (N-51) compared to 17α-ethynyl-17β-hydroxyestr-4-en-3-one (N-7) (by the same authors); and 17α,13β-diethyl-19β-hydroxygona-4,9(10)-dien-3-one (N-98) compared to 17α-ethyl-17β-hydroxyestra-4,9(10)-dien-3-one (N-97) all exhibit large increases in the androgenic and anabolic activities.

Birch [193] pointed out, as we have seen already, that the fact that a C-18 methyl group is necessary for the activity of steroids with a five-membered D ring may be due to its effect in preventing this junction from assuming the more stable cis configuration, when, for example, a C-17 carbonyl group is present at any intermediate stage at the synthesis. Since the trans junction is the more stable for two fused six-membered rings, the expectation is that with a six-membered ring D the angular C-18 methyl may be omitted with retention of activity. That removal of the C-18 methyl group in the D-homo series (E-23, E-32) did not produce an increase in the activity points to the fact that introduction of C-19 alkyl substituent can be accommodated on the β-side of the molecule. Thus introduction of C-19 methyl and in some cases C-19 ethyl substituent does not cause a decrease, but an increase in the activity.

Introduction of a bulky substituent at C-11 or C-8 decreases the androgenic and anabolic potencies; this points to the importance of β-face attachment both at C-19 and at C-18.

The size of an ethyl group is just about the limit, as is shown by the fact that 13β-ethyl-17β-hydroxy-7α-methylgon-4-en-3-one (N-52) compared to 17β-hydroxy-7α-methylestr-4-en-3-one (N-11) has decreases in both potencies and the anabolic–androgenic ratio as well.

We assume that β-face attachment is important at the C-13 position and that the steric requirements can accommodate an ethyl group at C-13. This substitution is tested only in the 19-nor series. We assume, therefore, that the β-face attachment is facilitated by a sloping plane extending from the 18-ethyl group (highest point through the β-face of C-8 and C-11) to the β-faces of C-6 and C-10 (which are equally distant from the 18-ethyl group).

14,15-SUBSTITUTION

Δ^{14}-Testosterone (S-63) and Δ^{14}-5α-dihydrotestosterone (D-48) show increased androgenic and anabolic activities in the rat when administered orally and decreased activities when administered subcutaneously. Chick comb assays for androgenic activity show that the Δ^{14}-compounds are more active than their parent compounds. However, in the mouse androgen assay Δ^{14}-5α-dihydrotestosterone is more active and Δ^{14}-testosterone is less active than their parent compounds.

These investigations were carried out [80] in the hope of altering the relation between the 17β-hydroxyl group and the rest of the molecule.

Because of the limited number of C-14 and C-15 compounds available, it is hard to generalize the effect of 14- and 15-substitution.

16-SUBSTITUTION

16,16-Difluorotestosterone (S-100) has no androgenic or anabolic activity. 16α-Methyltestosterone (S-129) and 16β-methyltestosterone (S-130) did not show increased anabolic or androgenic activities compared to testosterone. 16α,17α-Dimethyl-$\Delta^{9(11)}$-testosterone (S-131) and 16,17α-dimethyl-9α-fluoro-11β-hydroxytestosterone (S-132) showed decreases of both androgenic and anabolic activities to a great extent compared to 17α-methyl-9α-fluoro-11β-hydroxytestosterone (S-19). Both 16α-fluoro-16β-methyltestosterone (S-141) and 16α-fluoro-16β-methyl-5α-dihydrotestosterone (D-74) exhibited 20% of the androgenic activity of testosterone and no anabolic activity. 16-Hydroxymethylene-5α-dihydrotestosterone (D-23) was inactive as an androgen and anabolic agent. Both 4,4,6,16α-tetramethyl-3β-17β-dihydroxyandrost-5-en (A-69) and 4,4,6,16α-tetramethyl-17β-hydroxyandrost-5-en-3-one (A-82) possess 12% of the androgenic activity of testosterone and no anabolic activity. 16β-Methyl-19-nortestosterone (N-21) exhibits 10% androgenic and 40% anabolic activity of testosterone propionate.

Substitution at the 16-position was studied primarily because 16α-methyl substitution is known to increase anti-inflammatory activity in the cortical hormone series [180]. However, in the androgenic–anabolic series, both types of substitution decrease activity. At the same time, 16-hydroxymethylene-5α-dihydrotestosterone (D-23) abolishes activity. In this compound, the 16-substituent is no longer an α- or a β-substituent but projects in the plane of the molecule. This points to the importance of protrusion towards the third dimension. We will elaborate on this matter in the discussion of 17-substitution.

17-SUBSTITUTION

Without any doubt, the attachment of the steroid to the receptor at the 17-hydroxyl group is one of the most important attachments. This is also a point where most of the theories fail to explain the observed facts.

We have followed the principle of minimal structural requirement throughout this study. We selected 5α-androstane as the basic compound which displays the minimal structural requirements to bring about hormonal, specifically androgenic and anabolic, activity. It was demonstrated that this activity is a property of the hydrocarbon steroid skeleton of the natural androgen, 5α-androstane. However, introduction of an oxygen function enhances the activity to a large extent, particularly when the oxygen is introduced to the steroid skeleton as a 17β-hydroxyl group. Compare the high activity of 5α-androstan-17β-ol (D-145) to 5α-androstane. The presence of oxygen in the form of a 17β-hydroxyl group is more important than the presence of oxygen at carbon-3 and for practical purposes all useful anabolic agents and highly active androgens possess the 17β-hydroxyl group.

17α-Hydroxy Epimers

17α-Hydroxytestosterone (E-3), 1β,17β-dimethyl-17α-hydroxy-5α-androst-2-ene (E-4), 17α-iso-19-nortestosterone (E-19), 17β-methyl-17α-hydroxy-D-homoandrost-4-en-3-one (E-24), 3β,17α-dihydroxy-5α-androstane (E-34), 17β-methyl-17α-hydroxytestosterone (17-epitestosterone) (E-33), 3α,17aα-dihydroxy-5α-androstane (E-38), 17aα-hydroxy-D-homotestosterone (D-homo-*cis*-testosterone) (E-41), and 17α-hydroxy-13α-androst-4-en-one (E-54) derivatives possessing a 17α-hydroxy substituent have no androgenic and anabolic activity. This alone already points to the fact that the *hydroxyl group at C-17 is not in contact with the receptor surface at the α-face of ring D*, as was suggested by Wolff [71].

17-Ketone Compounds

Δ^4-Androsten-3,17-dione (S-22), 1α-cyano-Δ^4-androsten-3,17-dione (S-37), 1β-acetylthio-Δ^4-androstene-3,17-dione (S-38), 7α-methyl-Δ^4-androstene-3,17-dione (S-23), Δ^4-androstene-3,11,17-trione (S-43), 11β-hydroxy-Δ^4-androstene-3,17-dione (S-44), 3α-hydroxyandrostane-17-one (D-27), 3β-hydroxyandrostane-17-one (D-161), and androstane-3,17-dione (D-30) compared to the corresponding 17β-hydroxy derivatives exhibit much smaller androgenic and anabolic activities with unfavorable anabolic–androgenic ratios (with the exception of S-23, which has a favorable ratio).

We have already summarized the comparative activities of testosterone metabolites, many of which possess a 17-ketone function. Those results also clearly show the decreased androgenic potencies of 17-ketone metabolites.

The Role of the 17β-Hydroxyl Group

We have already considered the metabolic inactivation of the 17β-hydroxyl group and the effect of structural modification, namely introduction of a 17α-alkyl substituent, on the metabolism. At the present, no other type of substituent at C-17 is known which will replace the 17β-hydroxyl group in effectiveness. However, a large number of 17β-hydroxy esters, ethers, acetals, and ketals have been prepared which do not decrease the androgenic and anabolic activities. It might be argued that the C-17 esters are expected to hydrolyze first and the free 17β-alcohols are the active species. First findings suggested that the esters are not hydrolyzed; the intact esters are absorbed and take effect as such [178]. The effect of the ester group was believed to be due to a slowing down of the transport of the compound in the organism judging from the correlation between the rate of disappearance of the esters from the site of injection and the duration of action [178]. However, later reports substantiate the expectation that the free 17β-alcohols are the active species [381], produced by hydrolysis of the ester in vivo. In any event, regardless of whether the active species is the free alcohol or the intact ester, the rate-determining process will be the absorption of ester from an intramuscular depot. The rate of absorption of the ester can be measured by its disappearance from the site of injection. The duration of action and the ratio of the anabolic and androgenic activities will be determined by the rate of absorption of the esters from the intramuscular depot. The same considerations probably also hold for the other 17-hydroxy derivatives.

We conclude that the 17β-oxygen atom is responsible for attachment to the receptor site, whether in the form of a hydroxyl group or in the form of hydroxyl derivatives, such as esters, ethers, acetals, or ketals. We postulate that the attachment is a twofold attachment: the oxygen attaches to the receptor at the β-face and in addition in the third dimension, that is, to the plane which is perpendicular to the α- and β-faces of the steroid molecule. The necessity for the third-dimensional attachment is emphasized by the consideration of 13β-ethyl and $\Delta^{14,15}$-compounds. Inspection of 13β-ethyl-substituted derivatives, a very potent class of compounds, shows that the 13β-ethyl group severely interferes with a true β-face attachment of the 17β-hydroxyl group to the receptor. Therefore, some other form of attachment must also be operating. Inspections of the molecular model of $\Delta^{14,15}$-testosterone show that the skeleton is no longer "planar"; it is rather concave and the 17β-hydroxyl group is no longer a true β-substituent. While attachment on the β-side is not prevented the importance of third-dimensional attachment is also well documented by the C—O bond direction.

The Role of 17α-Alkyl Substituent

In Ringold's hypothesis [3] of α-face attachment of androgens to the receptor, one important point was that while replacement of the 17α-hydrogen by a methyl substituent maintains the high androgenic activity of the parent compound, the increasing length of the 17α-alkyl substituent (the change from methyl to ethyl) results in a marked decrease in androgenic activity. However, a large number of highly active 17α-ethyl compounds have been prepared since. Bush's [4] main objection to Ringold's theory was that β-sided association is suggested by the activity of epimeric 17-hydroxy steroids. Bush [7] in a later article favors a β-sided attachment of androgens to its receptors even though 17α-alkyl substituents larger than ethyl groups cause a sharp drop in androgenic and anabolic activity. According to Bush [7] "the problem may however be resolved by the extremely interesting analysis of Wolff, Ho, and Kwok [71] who suggest β-sided attachment of androgen over rings A, B and C and α-sided attachment to a second receptor area over ring D."

We have already discussed the studies of Klimstra et al. [87] dealing with the validity of Wolff's theory at Ring A.

Also we outlined a serious objection to Wolff's theory, when we considered the 17α-hydroxyandrostane epimers, namely, if Wolff's theory were correct and the hydroxyl group at C-17 were in contact with the receptor surface at the α-face, then surely the 17α-hydroxy epimers are in a much better position to attach themselves to the receptor surface at the α-face and should show considerably higher activity than the 17β-hydroxy epimers. But this is clearly not the case.

Furthermore, a large number of 17α-ethyl-17β-hydroxy–substituted compounds are highly active. Since the 17α-ethyl group is free to rotate and by steric considerations has an approximate van der Waals radius of 4 Å it can be seen that the 17β-hydroxyl group can attach itself to the receptor at the α-side only if the *binding forces penetrate through the area occupied by the freely rotating 17α-ethyl group.*
This is highly unlikely, to say the least.

It remains to consider the role of 17α-alkyl substituents in the attachment to the receptor. We can see that the steroid–receptor attachment can accommodate a 17α-methyl or 17α-ethyl group but there is a sharp decrease in activity when a 17α-propyl, or ethynyl group is introduced and the importance of the third-dimensional attachment becomes immediately evident. Inspection of molecular models reveals that a 17α-propyl or 17α-ethynyl substituent protrudes well beyond the molecular dimensions of steroids in the α and in the planar direction. In order to obtain the molecular dimensions of steroids we draw a plane perpendicular to the α- and β-faces, which connects along the equatorially oriented substituents at carbons 1, 2, 11, and 12 on one side, 16 and 17 on the other, and 6 and 7 on the third side. We postulate, therefore, that the steric requirements of the steroid–receptor attachment can accommodate 17α-methyl and 17α-ethyl substituents and the attachment of 17α-alkyl group is two-dimensional at the β-face and in the third dimension as defined above.

A NEW THEORY OF STEROID–RECEPTOR INTERACTION

After having examined the effects produced by alterations and substitutions at each carbon atom, we are in a position to postulate a new theory for the steroid hormone–receptor interaction.

In considering any kind of interaction between a steroid and receptor, we must remember that no matter how attractive it seems to picture the steroid skeleton as having two sides, an α-face and a β-face and no matter how many attempts there have been in trying to consider steroid–receptor interaction in terms of α- or β-side attachment, nevertheless one must not forget, that *the steroid is a three-dimensional entity.* When one tries to explain the behavior of a three-dimensional entity in terms of two dimensions, one gets a confusing picture. We postulate that the *steroid–receptor interaction is a three-dimensional attachment.* The steroid acts like a porcupine with shorter and longer quills and only some of the quills really get attached to the receptor. The receptor is flexible in accommodating a lot of the structural features of steroids.

The principal points of attachment can be described in terms of molecular dimensions in the following way:

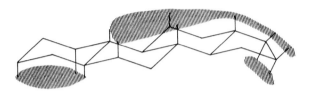

The important points of attachment are ring A on the α-side (through the axially oriented bonds attached at C-1, C-3, and C-5), rings B, C, and D on the β-side [through the axially oriented bonds attached at C-6, C-10, C-13, and secondarily C-8, C-11, and through the bonds attached at C-16 and C-17 (16β and 17β)]. In addition, on the α-side of ring D (through the bonds attached at C-16 and C-17). In addition, planar, peripheral (or p) attachment is made to planes perpendicular to the α- and β-sides

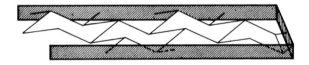

of the steroid skeleton. These planes are obtained along the equatorially oriented substituents of C-1, C-2, C-11, and C-12 on one side, C-16 and C-17 on the second side, and along the equatorially oriented substituents of C-6 and C-7 on the third side. The important points of attachment are ring A (through the equatorially oriented bonds attached at C-1 and C-2), ring B (through the equatorially oriented bonds attached at C-11 and C-12), ring C (through the equatorially oriented bonds attached at C-6 and C-7) and ring D (through the bonds attached at C-16 and C-17).

The interaction of androgens with a receptor necessary to elicit an androgenic and anabolic response *is on the α and p sides of ring A; β and p sides of ring B; β and p sides of ring C, and α,β- and p-sides of ring D.*

The points of attachment at C-8 and C-11 (β-side) are of secondary importance. They are responsible in certain conformational situations (for example, in the presence of 19-methyl and 18-methyl groups) for inhibiting the steroid–receptor attachments normally present at other positions by steric interference. Yet, in some other conformational situation (for example, 19-nor steroids), the absence of other bulky substituents render C-8 and carbon-11 positions on the β-face to become primary points of attachment to the receptor.

CHAPTER 4

THERAPEUTIC ACTION:

ANABOLIC–ANDROGENIC RATIOS

THE ANABOLIC–ANDROGENIC RATIO

For clinical use, anabolic steroids should lack all androgenic properties; however, such compounds have not been reported. For practical purposes the anabolic–androgenic ratio is used as a measure of the usefulness of anabolic agents.

The anabolic and androgenic activities are compared to a standard, usually testosterone or testosterone propionate (subcutaneous administration) or 17α-methyltestosterone (oral administration). There are four ways in which a favorable anabolic–androgenic ratio can be brought about:

(1) A large increase in the androgenic property coupled with an even larger increase in the anabolic activity.

(2) An increase in the anabolic activity while maintaining the androgenic activity in the vicinity of that of the standard.

(3) A decrease in the androgenic activity while maintaining the anabolic activity in the vicinity of that of the standard.

(4) A decrease in the androgenic activity coupled with an increase in the anabolic activity.

The last case approaches the ideal situation to the greatest extent.

Compounds Displaying Increased Androgenic Activity
Coupled with an Even Larger Increase
in the Anabolic Activity

9α-Fluoro-17α-methyl-11β,17β-dihydroxyandrost-4-en-3-one. This compound (Fluoxymesteron, S-19) was prepared in the hope of obtaining a large increase in the androgenic and anabolic activities similar to that observed in the corticoid series by 9α-fluoro 11β-hydroxy substitution. The compound was found to possess 9.5 times the androgenic and 20 times the anabolic activity of 17α-methyltestosterone [37]. No difference in the activities was found between the 11β-hydroxy and the 11-keto (S-25) derivatives. In spite of the fact that fluoxymesterone has a favorable anabolic–androgenic ratio, due to the strong androgenic effect the compound has been mainly used as an androgen, for example, as a highly active oral substitute for testosterone in hypogonadic conditions [267] or as a phallotropic androgen [268].

7α,17α-Dimethyl-17β-hydroxyandrost-4-en-3-one.   This 7α-methyl compound (Bolasterone, S-20) also exhibits a favorable anabolic–androgenic ratio but has a very strong androgenic effect. Introduction of a 7α-methyl substituent [269] into testosterone analogs usually leads to a great enhancement in activities. This was attributed to the fact that the 7α-methyl group inhibits the metabolism in the organism. The influence of the compound on the liver function has been reported [38].

17α-Methyl-17β-hydroxy-estra-4,9,11-trien-3-one   (Methyltrienolone, N-99).   In the course of a steroid total synthesis [293] this compound (Methyltrienolone, N-99) was prepared; it was reported [55] to possess 6000% of the androgenic (ventral prostate index), 7500% of the androgenic (seminal vesicles index), and 12,000% of the anabolic (levator ani index) activity of methyltestosterone. Later the anabolic activity was reported [74] to be 30,000% of that of methyltestosterone. As measured by multiple parameters methyltrienolone turned out to be the most "hepatotoxic steroid," causing biochemical symptoms of intrahepatic cholestasis [75]. It was also reported [74] to reduce the excretion of 17-ketosteroids and 17-hydroxycorticosteroids and to cause enhancement of the blood coagulation factors V, VII, and X. It also increases the prothrombin content of the plasma [74].

Others.   In addition a number of 7α-methyl-substituted 19-nortestosterone derivatives were found to possess favorable anabolic–androgenic ratios with very high androgenic activities: 7α-methyl-17β-acetoxy-19-norandrost-4-en-3-one (N-9), $7\alpha,17\alpha$-dimethyl-19-norandrost-4-en-3-one (N-10), 7α-methyl-17β-hydroxy-19-norandrost-4-en-3-one (N-11), and 7α-methyl-19-norandrost-4-en-3,17-dione (N-18). The 7α-methylation is regarded as a protection of the 19-nor configuration against metabolic inactivation. The speculation was also put forward [35] that testosterone derivatives may owe their activities at the end organs (seminal vesicles, ventral prostate, and levator ani indices) to their transformation to 19-nortestosterone derivatives, the 19-nortestosterone derivatives being the active species.

In the 5α-androstane series introduction of an oxygen function into the molecule at carbon-1 caused a great increase in the androgenic and the anabolic activities with a resulting favorable anabolic–androgenic ratio; 17α-methyl-17β-hydroxy-5α-androstan-1-one (D-56) and 17α-methyl-1α,17β-dihydroxy-5α-androstane (D-59) have a ratio of 4.5 and 4.1, respectively, when administered orally.

Ether formation often leads to enhanced oral activity. 17β-Hydroxy-5α-androstan-3-one 17-(1'methoxy)cyclopentyl ether (D-130), 17β-hydroxy-5α-androstan-3-one 17-(1'-ethoxy)cyclopentyl ether (D-132), 17β-hydroxy-5α-androst-1-en-3-one 17-(1'-ethoxy)cyclopentyl ether (A-163), and 17β-hydroxy-5α-androst-1-en-3-one 17-cyclopent-1'-enyl ether (A-164) are all reported to possess favorable anabolic–androgenic ratios with increased anabolic and androgenic properties. The enhancement of oral activity is interesting in view of the fact that the ether bond of these compounds is easily broken, as manifested by the strongly enhanced excretion of 17-ketosteroids in humans treated with the ether derivatives [144].

In the 19-nortestosterone series introduction of a hydroxyl group leads to a substantial increase in the androgenic and anabolic potencies in two cases. 17α-Methyl-11β,17β-dihydroxy-19-norandrost-4-en-3-one (N-66) and 17α-methyl-4,17β-dihydroxy-19-norandrost-4-en-3-one (N-74) both display favorable anabolic–androgenic ratios. The increased activity of the 4-hydroxy derivative N-74 is particularly interesting, since 4-hydroxy substitution usually decreases the androgenic activity (see N-77).

Extension of unsaturation also increases the androgenic and anabolic activities in some cases. The heteroannular dienone, 13β,17α-diethyl-17β-hydroxygona-4,9(10)-dien-3-one (N-98) and several heteroannular trienones, 7α,17α-dimethyl-17β-hydroxyestra-4,9(10),11-trien-3-one (N-102), 17β-acetoxyestra-4,9(10),11-trien-3-one (N-103), 17β-methoxy-methyloxyestra-4,9,11-trien-3-one (N-104), 17β-hydroxyestra-4,9,11-trien-3-one-17-n-decanoate (N-105), and 17β-hydroxyestra-4,9,11-trien-3-one 17-carbobenzoxyate (N-106) all exhibit very large increases in both androgenic and anabolic activities. The ratio is not favorable.

COMPOUNDS DISPLAYING INCREASED ANABOLIC ACTIVITY WITH UNCHANGED ANDROGENIC ACTIVITY COMPARED TO THE STANDARD

17β-Hydroxy-5α-androstan-3-one.   This is a compound (Dihydro-testosterone, androstanolon, stanolon) (D-1) of short-lived activity and it has little advantage over testosterone propionate. The growth of the ventral prostate is stimulated to the largest extent, while the levator ani and seminal vesicle indices are slightly increased. The anabolic–andro-genic ratio is between 1 and 2.

17α-Methyl-17β-hydroxy-5α-androstan-3-one.   In oral administration this steroid (Methyldihydrotestosterone, mestanolon, D-2) shows similar properties to D-1.

17α-Methyl-3β,17β-dihydroxyandrost-5-ene.   This was the first compound (Methylandrostenediol, methandriol, A-5) used as an orally active anabolic agent with little androgenic activity. However, it was found later that the anabolic potency is also weaker than that of methyltestosterone [32, 245]. It is also used in the forms with longer lasting activity, such as dipropionate, 3-propionate, and 17-enanthoyl acetate.

2α,17α-Dimethyl-17β-hydroxy-5α-androstan-3,3′-azin.   This derivative (Dimethazin, D-5) was found to possess the same androgenic activity as methyltestosterone with a twofold increase of the anabolic activity. The anabolic–androgenic ratio is 2.

4,17β-Dihydroxy-17α-methyl-androst-4-en-3-one.   An increased anabolic effect and a slightly decreased androgenic effect are displayed by this compound (Oxymesterone, S-16). The change in biological activity achieved at oral administration may be due to the fact that the C-4 hydroxyl group facilitates intestinal absorption [127]. The compound shows a good anabolic–androgenic ratio (7.0) and good nitrogen retention in clinical tests [127].

1α,7α-Bis(acetylthio)-17α-methyl-17β-hydroxyandrost-4-en-3-one.   In comparison with 17α-methyltesterone, this compound (Thiomesterone, S-61) showed an increase in the anabolic activity and a slight decrease in the androgenic activity. The anabolic–androgenic ratio is favorable (7.5). Clinical investigations were also favorable with only slight side effects when administered orally [76]. The compound was not estrogenic, was barely active in progestation, and inhibited gonadotropin production, although to a lesser extent than methyltestosterone. The influence on the liver function has also been reported [38].

2-Hydroxymethylene-17α-methyl-17β-hydroxy-5α-androstane.   This derivative (Oxymetholon, D-10) was subjected to extensive investigations. In the preliminary report, the compound was found to be a potent orally active anabolic agent with a low androgenic activity [23]. These results were confirmed [78] a year later. Camerino and Sala [9] found the compound to have 150% of the anabolic and 30% of the androgenic; Dorfman and Kincl [30], 320% of the anabolic and 45% of the androgenic (orally), and 64% of the androgenic (subcutaneous) potency of methyltestosterone. Suchwosky and Junkmann [266] gave an anabolic–androgenic ratio of 2.5 (oral) and 6.0 (subcutaneous administration), but found that at the dose levels required to display a good anabolic effect, the compound shows considerable side effects (hypophysis inhibition) (anti-ICSH-effect). Edgren [81] found that the compound was markedly androgenic at doses that produced no acceleration of body weight over a 2-week period (at dose levels without anabolic response).

Others. In addition to the above-described commercially available compounds, a large number of compounds display favorable anabolic–androgenic ratios due to increased anabolic activity coupled with unchanged androgenic activity compared to the standard. Following is a list of these compounds:

1α-Methylthio-17α-methyl-17β-hydroxyandrost-4-en-3-one (S-46), 6α-chloro-17β-acetoxyandrost-4-en-3-one (S-29), 2,17α-dimethyl-17β-hydroxyandrost-1,4,6-trien-3-one (S-69), 7α,17α-dimethyl-17β-hydroxyandrost-1,4-dien-3-one (S-138), 9α,11β-dichloro-17α-methyl-17β-hydroxyandrost-1,4-dien-3-one (S-127), 17α-methyl-11β, 17β-dihydroxyandrost-4-en-3-one (S-18), 1α,7α-dimethyl-17β-acetoxy-androst-4-en-3-one (S-80), and 1α,7α-17α-trimethyl-17β-hydroxy-androst-4-en-3-one (S-94) among the testosterone analogs.

1α,17β-Dihydroxy-5α-androstan-3-one (D-89), 1α-hydroxy-17β-acetoxy-5α-androstan-3-one (D-107), 1α,17β-diacetoxy-5α-androstan-3-one (D-108), 1α,7α-dimethyl-17β-hydroxy-5α-androstan-3-one (D-96), 1α,7α,17α-trimethyl, 17β-hydroxy-5α-androstan-3-one (D-128), 2α,17α-dimethyl-17β-propionoxy-5α-androstan-3-one (D-6), 2α,6α,17α-tri-methyl-17β-propionoxy-5α-androstan-3-one (D-7), 2α-methyl-17β-hydroxy-5α-androstane-17-tetrahydropyranyl ether (D-35), 2α-methyl-17β-dichloroacetoxy-5α-androstan-3-one (D-158), 2α,17α-dimethyl-17β-hydroxy-5α-androst-9(11)-en-3-one (D-55), 2α-methyl-17β-hy-droxy-5α-androst-9(11)-en-3-one (D-50), 2-methylene-17α-methyl-17β-hydroxy-5α-androstane (D-9), 2α,17α-dimethyl-17β-hydroxy-5α-andro-stan-3-hydrazone (D-15), 2α,3α-difluoromethylene-17α-methyl-17β-hydroxy-5α-androstane (D-52), 2α,3α-epithio-17α-methyl-17β-hy-droxy-5α-androstane (D-70), and 17α-methyl-17β-hydroxy-5α-andro-stano[2,3-c]furazan (D-71) among the dihydrotestosterone analogs.

1α,17α-Dimethyl-17β-hydroxy-5α-androst-2-ene (A-72), 17β-hydroxy-5α-androst-1-en-3-one (A-7), 17α-methyl-17β-hydroxy-5α-androst-1-en-3-one (A-8), 17β-acetoxy-5α-androst-1-en-3-one (A-78), 2,17α-dimethyl-17β-hydroxy-5α-androst-1-en-3-one (A-11), 2-methyl-17β-acetoxy-5α-androst-1-en-3-one (A-61), 2-methyl-17β-hydroxy-5α-androst-1-en-3-one (A-10), 2-formyl-17α-methyl-17β-hydroxy-5α-androst-1-en-3-one (A-63), 17α-methyl-17β-hydroxy-5α-androst-2-ene (A-22), 2,17α-dimethyl-17β-methyl-17β-hydroxy-5α-androst-2-ene (A-22), 2,17α-dimethyl-17β-hydroxy-5α-androst-2-ene (A-24), 2-cyano-17β-caproyloxy-5α-androst-2-ene (A-27), 2-cyano-17α-methyl-17β-hydroxy-5α-androst-2-ene (A-43), 2-cyano-17β-hydroxy-5α-androst-2-ene (A-44), 2-hydroxymethyl-

17β-hydroxy-5α-androst-2-ene (A-38), and 2-hydroxymethyl-17α-methyl-17β-hydroxy-5α-androst-2-ene (A-26), among the 5α-androstene derivatives containing ring A unsaturation.

COMPOUNDS DISPLAYING DECREASED ANDROGENIC ACTIVITY WITH UNCHANGED ANABOLIC ACTIVITY COMPARED TO THE STANDARD

19-Nortestosterone Derivatives

Hershberger et al. [187] discovered in 1953 that although 19-nortestosterone and other 19-nor analogs exhibit only relatively weak androgenic properties the protein anabolic effects and myotrophic effects were of the same order of magnitude as those of testosterone. Since the anabolic–androgenic ratio of the 19-nortestosterone derivatives was very favorable, this discovery opened a new field of anabolic agents.

19-Nortestosterone (N-1) and 19-Nortestosterone Esters. The original results (3.33 anabolic–androgenic ratio) of Hershberger [187] were confirmed a year later [65]. In spite of the favorable ratio, nortestosterone has only short-lived activity and therefore the 17-esters are used clinically, preventing quick metabolic inactivation of the 17β-hydroxyl group.

19-Nortestosterone phenylpropionate was found [246] to have a myotrophic effect four times stronger and an androgenic effect only half that of testosterone phenylpropionate. Nitrogen and calcium retention values were also favorable [238].

19-Nortestosterone cyclopentylpropionate has a myotrophic action approximately twice that of testosterone propionate but only about 20% of the androgenic properties of testosterone propionate [65]. In addition, 19-nortestosterone cyclohexylpropionate [247], 19-nortestosterone *n*-decanoate [148,248], 19-nortestosterone adamantoate [147], 19-nortestosterone 17β-*p*-hexoxyphenyl propionate [249], 19-nortestosterone 17β-hemisuccinate (sodium salt), and 19-nortestosterone propionate are in clinical use.

17α-Ethyl-17β-hydroxy-19-norandrost-4-en-3-one. A series of 17α-alkyl-19-nortestosterone derivatives were prepared for clinical evaluation. The first member of the series, the 17α-methyl compound (N-4), showed in addition to good anabolic activity a substantial androgenic and strong progestational and antigonadotrophic activities. Clinically it is used as an oral gestagen. The 17α-ethyl derivative (Ethylnortestosterone, Norethandrolon, N-5), however, exhibits such a strong anabolic activity and favorable anabolic–androgenic ratio (3–5) that in spite of its progestational [252], deciduomatogenic [253], antiestrogenic [254], and antigonadotrophic activity [255] it is widely used as an anabolic agent.

Nitrogen retention evaluation [251] was as good with 17α-ethylnortestosterone as with testosterone propionate (subcutaneous administration) or 17α-methyltestosterone (oral administration) [250]. The calcium and phosphorus retention was also favorably affected but as a side effect, mild cholestatic jaundice was observed [256]. Also a marked increase in bromsulfalein retention was observed in humans by administration of 17α-ethylnortestosterone [257]. Other 17α-alkyl substitutions (propyl, allyl, methallyl, ethynyl) caused a decrease of the anabolic activity. A new type of derivatives of 19-nortestosterone was prepared by etherification of the 17β-hydroxy group [12, 143, 258]. Among these compounds, 19-nortestosterone-17-(cyclopent-1'-enyl) ether (N-81) possesses a favorable anabolic–androgenic ratio without the undesirable side effects.

4,17β-Dihydroxy-19-norandrost-4-en-3-one 17-cyclopentylpropionate. This compound (Oxabolon, N-77) has prolonged anabolic activity and reduced androgenic properties [244], to bring about a favorable anabolic–androgenic ratio. The undesirable side effects found at some 19-nortestosterone derivatives are reduced [127] by introduction of the 4-hydroxyl group. In contrast, 4-hydroxy-17α-methyl-19-nortestosterone (N-74) shows a strong increase of the anabolic and the androgenic effects with a moderate increment [127] of the therapeutic index as compared with 17α-methyl-19-nortestosterone (N-4). The anabolic–androgenic ratio compared to 17α-methyltestosterone still remains favorable (4.5).

4-Chloro-17β-acetoxy-19-norandrost-4-en-3-one. It was reported that this derivative (N-14) was [126] slightly less anabolic (80%) and androgenic (50%) than 17α-ethylnortestosterone. Significant inhibition of pituitary gonadotropins was also manifested. The compound was also found active in retaining nitrogen in castrated male rats submitted to balance studies [259].

Others. The 19-nortestosterone analogs, which instead of the Δ^4–3-ketone system contain the Δ^5–3-ketone system (unconjugated ketone) showed very similar androgenic and anabolic potencies compared to the Δ^4–3-ketone compounds [68]. It is most likely that the Δ^5–3-ketone compounds, 17α-ethynyl-17β-hydroxy-19-norandrost-5-en-3-one (N-37), 17α-ethyl-17β-hydroxy-19-norandrost-5-en-3-one (N-38), and 17α-methyl-17β-hydroxy-19-norandrost-5-en-3-one (N-39) under the influence of the acid pH of the stomach are rearranged to the Δ^4–3-ketones.

However, the compounds containing the Δ^5–3β-ol system where rearrangement due to the lack of 3-ketone group is no longer likely to happen, show highly favorable anabolic–androgenic ratios. Thus, 17α-methyl-$3\beta,17\beta$-dihydroxy-19-norandrost-5-ene (N-26) and 17α-ethyl-$3\beta,17\beta$-dihydroxy-19-norandrost-5-ene (N-27) exhibit very favorable anabolic–androgenic ratios (14 and 20, respectively). A large number of 19-nortestosterone analogs have been prepared containing a $\Delta^{5(10)}$–3-ketone system instead of the Δ^4–3-ketone system. Unconjugation of the double bond in this direction leads to unfavorable changes in the androgenic and anabolic activities. While 17α-methyl-17β-hydroxy-19-norandrost-5(10)-en-3-one (N-40) still exhibits a favorable anabolic–androgenic ratio with decreased activities, the other analogs, 17α-ethynyl-17β-hydroxy-19-norandrost-5(10)-en-3-one (N-43), 17β-hydroxy-19-norandrost-5(10)-en-3-one (N-92), and 17α-ethyl-17β-hydroxy-19-norandrost-5(10)-en-3-one (N-93) show sharply reduced activities. Ether formation at the 17β-hydroxyl position helped to regain a favorable anabolic–androgenic ratio, i.e., 17β-hydroxy-19-norandrost-5(10)-ene 17-(cyclopent-1'-enyl) ether (N-71) has a ratio of 3 and 17α-methyl-17β-hydroxy-19-norandrost-5(10)-ene 17-(cyclopent-1'-enyl) ether (N-83) has a ratio of 2.2.

In addition to the above-described compounds, there are a few other 19-nortestosterone derivatives which display a favorable anabolic–androgenic ratio by having lower androgenic activities and maintaining the anabolic activity of the standard compound: 17α-methyl-17β-hydroxy-19-norandrost-2-ene (N-29), 2-cyano-17β-acetoxy-19-norandrost-2-ene acetoxy-19-norandrost-4-eno[2,3-d]isoxazole (N-70), 17α-methyl-17β-hydroxy-19-norandrost-4-eno[2,3-d]isoxazole (N-76), and 17α-methyl-17β-hydroxy-5α-estran-3-one (N-90).

Androstane Derivatives

4-Chloro-17β-acetoxy-androst-4-en-3-one. Extensive studies at different dosages [243] showed this compound (Chlorotestosterone acetate, S-146) to have an anabolic–androgenic ratio of 5 and 70% of the anabolic activity of testosterone propionate. It caused an increase of body weight at subcutaneous administration in female adult rats [243]. Balance studies in humans showed that 4-chlorotestosterone acetate showed nitrogen retention [260–262]. Chlorotestosterone acetate has no progestational estrogenic or corticoid properties; the antiestrogenic property is 33% of that of testosterone propionate [243]. The presence of a chlorine atom at C-4 slows down considerably the oxidation of the 17β-hydroxyl group to a 17-keto group [239] and chlorotestosterone acetate was also shown to be inert towards ring A aromatization [234].

All other 17-ester derivatives — 4-chlorotestosterone 17-propionate (S-6), 17-succinate, 17-palmitate, 17-enantate, 17-phenylpropionate, 17-triphenylmethyl ether, 17-phenoxyacetate, and 17-cyclopentylpropionate [243] — were reported to be less active than 4-chlorotestosterone acetate. However, 4-chlorotestosterone chloroacetate was reported [263] to be several times more active than 4-chlorotestosterone acetate and 4-chlorotestosterone *p*-chlorophenoxyacetate was promising in showing a long lasting effect [234, 263].

4-Chloro-17α-methyl-17β-hydroxyandrost-1,4-dien-3-one. This derivative (S-105) was reported to have a much higher anabolic–androgenic ratio than Δ^1-17α-methyltestosterone [263]. It was shown that application of S-105 does not increase the estrogen excretion [234].

17α-Methyl-17β-hydroxyandrost-1,4-dien-3-one. Preparation of this compound (Methandrostenolon, Methandienone, 1-Dehydromethyl-testosterone, S-15) was a result of investigations on the effect of additional unsaturation in testosterone derivatives. It was found to show strong anabolic properties and low androgenic activity [264]. The nitrogen retention values were favorable and no progestational side effects were demonstrated [264]. It is expected that the aromatization of ring A proceeds at a much lower rate than that of 17β-hydroxyandrost-1,4-dien-3-one due to the rate-decreasing effect of the 17α-methyl group. While comparative studies in this case are not available, it was shown that administration of 17α-methyl-17β-hydroxyandrost-1,4-dien-3-one contributes to the increase of the estriol fraction in the urine [265].

Androsta-1,4-dien-17β-ol-3-one 17-(cyclopent-1 '-enyl) ether. The 17-ether function instead of the 17α-methyl group in this derivative (Quinbolin, S-147) [14] is present in order to assure protection against metabolic inactivation of the 17-hydroxyl group [12]. The compound is orally applied and shows androgenic and anabolic activities similar to those of 17α-methyl-17β-hydroxyandrost-1,4-dien-3-one.

2α-Methyl-17β-hydroxy-5α-androstan-3-one. This compound (Dromostanolone, Drostanolon, D-49) was found to possess 130% of the anabolic and 40% of the androgenic activity of testosterone, therefore exhibiting a good anabolic–androgenic ratio (3). Esterification of the 17β-hydroxy group by propionic acid (D-28) increased the anabolic activity to 166% of that of testosterone. Even more remarkable was the effect of the etherification of the 17β-hydroxyl group: the 2'-tetrahydropyranyl ether of 2α-methyldihydrotestostcrone (D-35) showed 88% (ventral prostate), 82% (seminal vesicles), and 395% (levator ani) of the activity of methyltestosterone by oral administration [146].

Other. In addition to the commercially available compounds which have been discussed, the following other derivatives exhibit a favorable anabolic–androgenic ratio by having a lower androgenic activity but the same anabolic activity as the standard compound:

1α-Methyl-17β-acetoxyandrost-4-en-3-one (S-72), 1β-methyl-17β-acetoxyandrost-4-en-3-one (S-73), 1α-ethylthio-17α-methyl-17β-hydroxyandrost-4-en-3-one (S-47), 1α-acetylthio-17α-methyl-17β-hydroxyandrost-4-en-3-one (S-48), 1α-methyl-4-chloro-17β-acetoxyandrost-4-en-3-one (S-79), 1α,2α-methylene-17β-acetoxyandrost-4-en-3-one (S-78), 1α,2α-methylene-11β-hydroxy-17β-acetoxyandrost-4-en-3-one (S-83), 1α,7α-diethylthio-17α-methyl-17β-hydroxyandrost-4-en-3-one (S-59), 17β-hydroxyandrost-1,4-dien-3-one (S-70), 4-chloro-17β-acetoxyandrost-1,4-dien-3-one (S-114), 9α-bromo-11β-fluoro-17β-propionoxyandrost-1,4-dien-3-one (S-124), 2,17α-dimethyl-17β-hydroxyandrost-1,4-dien-3-one (S-68), 2-methylene-17α-methyl-17β-hydroxyandrost-4-en-3-one (S-32), 3-methylene-17α-methyl-17β-hydroxyandrost-4-ene (S-26), 4-methyl-17β-hydroxyandrost-4-en-3-one (S-11), 4-chloro-11β-hydroxy-17β-acetoxyandrost-4-en-3-one (S-7), 17α-methyl-17β-hydroxyandrost-4,6-dien-3-one (S-107), 6-chloro-17β-acetoxyandrost-4,6-dien-3-one (S-64), 7α-mercapto-17α-methyl-17β-hydroxyandrost-4-en-3-one (S-53), and 7α-ethylthio-17α-methyl-17β-hydroxyandrost-4-en-3-one (S-55), compounds in the testosterone series.

1α-methyl-17β-hydroxy-5α-androstan-3-one (D-84), 1α-methyl-17β-acetoxy-5α-androstan-3-one (D-100), 1α,17α-dimethyl-17β-hydroxy-5α-androstan-3-one (D-123), 1α,2α-methylene-17β-acetoxy-5α-androstan-3-one (D-112), 2α,17α-dimethyl-17β-hydroxy-5α-androstan-3-one (D-45), 2α-hydroxymethyl-17β-hydroxy-5α-androstan-3-one (D-8), 2-aminomethylene-17α-methyl-17β-hydroxy-5α-androstan-3-one (D-78), 2-[2′-(*N,N*-dimethylamino)ethylaminomethylene]-17α-methyl-17β-hydroxy-5α-androstan-3-one (D-79), 2-[2′-(*N,N*-diethylamino)ethylaminomethylene]-17α-methyl-17β-hydroxy-5α-androstan-3-one (D-80), 2-*N,N*-diethylaminomethylene-17α-methyl-17β-hydroxy-5α-androstan-3-one (D-83), 2-*N,N*-dimethylaminomethylene-17α-methyl-17β-hydroxy-5α-androstan-3-one (D-82), 2α,3α-methylene-17β-hydroxy-5α-androstane (D-40), 2α,3α-epithio-17β-hydroxy-5α-androstane (D-69), 2β,3β-epithio-17α-methyl-17β-hydroxy-5α-androstane (D-66), 2α,3α-difluoromethylene-17β-acetoxy-5α-androstan-3-one (D-51), ζ-isothiocyano-17β-hydroxy-5α-androstan-3-one (D-90), 4β-methyl-17β-hydroxy-5α-androstan-3-one (D-46), and 4α-methyl-17β-hydroxy-5α-androstan-3-one (D-47) in the 5α-dihydrotestosterone series.

1α,17β-Dihydroxy-17α-methyl-5α-androst-2-ene (A-70), 17α-methyl-17β-hydroxy-5α-androst-2-en-1-one (A-67), 17β-acetoxy-5α-androst-1,3-diene (A-45), 17β-hydroxy-5α-androst-2,4-diene (A-153), 2-methyl-17β-hydroxy-5α-androst-2-ene (A-23), 1β,17α-dimethyl-17β-hydroxy-5α-androst-2-ene (A-127), 2-formyl-17α-methyl-17β-hydroxy-5α-androst-2-ene (A-28), 17α-methyl-17β-hydroxy-2-bromo-5α-androst-1-en-3-one (A-74), 17α-methyl-17β-hydroxy-2-chloro-5α-androst-1-en-3-one (A-75), 2-formyl-17β-hydroxy-5α-androst-2-ene (A-36), 2-difluoromethyl-17β-acetoxy-5α-androst-2-ene (A-40), 2-fluoromethyl-17β-acetoxy-5α-androst-2-ene (A-41), 2-fluoromethyl-17β-hydroxy-5α-androst-2-ene (A-64), 17α-methyl-17β-hydroxy-5α-androst-2-en-4-one (A-68), 17β-acetoxy-5α-androst-2-en-4-one (A-81), 17α-methyl-17β-hydroxy-5α-androstano[3,2-d]-2′-methylpyrimidine (A-84), 6β,17α-dimethyl-17β-hydroxy-5α-androst-2-ene (A-71), 17α-methyl-17β-hydroxyandrost-4-eno[2,3-d]isoxazole (A-58), 17β-acetoxyandrost-4-eno[2,3-d]isoxazole (A-147), 17β-hydroxyandrost-4-eno[2,3-d]isoxazole (A-149), 17α-methyl-17β-hydroxyandrost-4-eno[3,2-c]pyrazole (A-157), and 5α-androst-1-en-17β-ol (A-1) among the 5α-androstene derivatives containing ring A unsaturation.

COMPOUNDS DISPLAYING DECREASED ANDROGENIC ACTIVITY
COUPLED WITH INCREASED ANABOLIC ACTIVITY

17α-Methyl-17β-hydroxy-2oxa-5α-androstan-3-one. Upon oral administration this derivative (Oxandrolone, E-21) was found to possess 322% of the anabolic and 24% of the androgenic activity of methyltestosterone. Oxandrolone caused nitrogen retention at all dose levels as evidenced by significant reductions in urinary nitrogen excretion during the test period [52]. Oxandrolone exhibited about 10% the potency of norethandrolone (N-5, 17α-ethyl-19-nortestosterone) as an inhibitor of pituitary gonadotropin. With an anabolic–androgenic ratio of 13, this compound is one of the most potent anabolic agents. This compound also represents the first example of a therapeutically useful steroid having a heteroatom (oxygen) inserted into the steroid nucleus.

17-Substituted 1-methyl-5α-androst-1-en-3-one. 1-Methyl-17β-hydroxy-5α-androst-1-en-3-one (Methenolone, A-86), 1-methyl-17β-acetoxy-5α-androst-1-en-3-one (Methenolone acetate, A-6), and 1-methyl-17β-hydroxy-5α-androst-1-en-3-one 17-oenanthate (Methenolone-oenanthate) compounds were the first compounds introduced [277, 278] among the anabolic agents which do not have a 17α-alkyl substituent and still possess a substantial oral activity compared to 17α-methyltestosterone. The side effects normally associated with orally active compounds possessing a 17α-alkyl group, namely, disturbance of the liver function, are absent; only slight increases in bromsulfalein and blood coagulation factors were observed [271, 302]. Methenolone caused significant nitrogent retention [270]. The significant oral anabolic activity of methenolone in the absence of a 17α-alkyl group or a 17-ester group may be due to the presence of the 1-methyl group, which appears to have an effect on the metabolic inactivation of the steroid: the oxidation of the 17β-hydroxyl group occurs to a lesser degree than the oxidation of testosterone by a liver enzyme preparation [270] and methenolone is not aromatized to estrogens [270]. Methenolone was extensively studied for growth accelerating effects [272]; absorption, distribution [273], excretion [274] metabolism [275] and clinical investigation [276] were favorable. Methenolone-oenanthate is used as a long lasting compound. The anabolic–androgenic ratio for methenolone acetate was reported to be 13 at oral administration and 16 at subcutaneous administration [45]. 17α-Methyl-17β-hydroxy-5α-androst-1-en-3-one (A-104) is also reported [100] to possess a favorable anabolic–androgenic ratio.

17α-Ethyl-17β-hydroxyestrene. This derivative (17α-ethyl-17β-hydroxy-19-norandrost-4-ene, Ethylestrenol, N-8) belongs to the 17-alkyl-3-desoxo-19-nortestosterone class of compounds [279]. It exhibits 420% of the anabolic activity and 22% of the androgenic activity of 17α-methyltestosterone [280] at oral administration. The anabolic–androgenic ratio is 19. A later publication [45] gave a substantially lower value for the anabolic–androgenic ratio (4.0 for oral, 3.7 for subcutaneous administration). In the following year, another group of workers reported an intermediate value of 8 for the anabolic–androgenic ratio at oral administration [63], nitrogen-retaining activity being taken as a basis for the evaluation of the anabolic activity. The clinical-pharmacological evaluation [281] confirmed the favorable preliminary results.

17α-Methyl-17β-hydroxy-5α-androstano[3,2-c]pyrazole. This compound (Stanozolol, Stanazol, Androstanazol, D-11) was found to be the most potent member of the steroidal [3,2-c]pyrazole series [203]. The compound was reported to possess 220% of the myotrophic activity and 33% of the androgenic activity of 17α-methyltestosterone [124] while nitrogen retention values in rats indicated that stanozolol at oral administration showed 25–45 times the activity of methyltestosterone. However, the anabolic activity evaluated on the basis of nitrogen retention in monkeys was not favorable [72]. Two further publications [26, 50] gave anabolic–androgenic ratios of 10.5 and 3.8 on the basis of levator ani evaluation, leaving the original [124] anabolic–androgenic ratio of 6 as an intermediate figure. The clinical evaluation of the anabolic activity was also positive [282].

17α-Methyl-17β-hydroxy-5α-androstan[3,2-c]isoxazole. Evaluation of this derivative (Androisoxazole, D-33) for anabolic activity by means of nitrogen balance studies gave an oral anabolic activity of 155% of that of 17α-methyltestosterone. At the same time, the androgenic activity (oral administration) was found to be 22% of that of 17α-methyltestosterone on the basis of weight increase of the ventral prostate. In spite of the fact that this compound has found therapeutical application [131, 283–285], it soon became apparent that the steroidal [2,3-d]isoxazoles possessed more noteworthy endrocinological activities [11] than the steroidal [3,2-c]isoxazoles.

17α-Methyl-17β-hydroxy-5α-androstan[2,3-d]isoxazole. The myotrophic activity of this compound (A-30) (based on the weight response of the levator ani muscle) is reported to be 200% that of 17α-methyl testosterone [49] and the androgenic activity (based on the increase of weight of the ventral prostate) to be 24% that of 17α-methyltestosterone when administered orally. However, based on nitrogen retention studies [49, 286, 287], methylandrostanolisoxazole (A-30) was reported to be 9.7 times as effective as methyltestosterone. (Compare 970% activity of A-30 to 155% activity of D-33 for the isomeric steroidal [2,3-d]- and [3,2-c]isoxazole, respectively.) Although introduction of an additional unsaturation between carbons 4 and 5 (A-58) still results in a favorable anabolic–androgenic ratio, the anabolic potency, as evaluated by myotrophic activity, is only of the same order of magnitude as that of 17α-methyltestosterone [11]. The anabolic potency of A-58 is 170% on the basis of nitrogen retention evaluation with a corresponding 21% androgenic activity on the basis of weight increase of the ventral prostate, compared to those of 100% for 17α-methyltestosterone [63]. Extension of the isoxazole series to higher homologs with additional unsaturation (A-148) or with 17α-alkyl groups larger than methyl resulted in a decline of activity [11].

17α,13β-Diethyl-17β-hydroxygon-4-en-3-one. This derivative (Norbolethone, N-46) was reported to possess 340% of the anabolic (myotrophic) and 17% of the androgenic (ventral prostate index) potency of testosterone propionate [81] when administered subcutaneously. The compound was also found to be active at oral administration [288, 289]. The anabolic–androgenic ratio at subcutaneous administration was found to be 20 on the basis of myotropic evaluation. Nitrogen retention and body weight effects were also favorable [81]. The oral anabolic activity in a nitrogen retention assay was 16.3 times that of methyltestosterone [290] and 4.2 times more active than stanozolol (D-11). The clinical evaluation [291] of the compound was also satisfactory. The compound was found to block the uterine growth effects of both estrone and estriol in mice and to have minor metrotropic effects when tested alone [289]. Since this compound has been obtained by total synthesis [292] the commercially available compound is a racemic mixture of *d* and *l* enantiomers. The 7α-methyl homolog, 7α-methyl-13β-ethyl-17β-hydroxy-gon-4-en-3-one (N-52), also exhibits a favorable anabolic–androgenic ratio (5).

17α-Methyl-17β-hydroxy-5α-androstane-2'-methyl[3,2-d]thiazole. This compound (A-85, Syntex Corp.) was reported [94] to be an orally effective anabolic agent showing 200% of the anabolic activity and 40% of the androgenic activity of methyltestosterone, when administered orally.

3-Methylene-17α-methyl-17β-hydroxy-5α-androst-1-ene. Upon oral administration this compound (A-21), (E. Merck A.G., Darmstadt, Germany) showed [48] 266% of the anabolic activity and 12% of the androgenic activity of methyltestosterone. The anabolic–androgenic ratio of 22 represents a very favorable therapeutic index.

17α-Ethyl-17β-hydroxyestra-4,9(10)-dien-3-one. This compound (Ethyldienolone, N-97, Eli Lilly & Co.) was reported [154*] to possess 30% of the androgenic and 150% of the anabolic potency of methyltestosterone when administered orally and 10% of the androgenic and 100% of the anabolic potency of methyltestosterone when administered subcutaneously. The corresponding 17α-methyl derivative (N-96) possesses higher activities, but a somewhat less favorable anabolic–androgenic ratio.

3,3-Azo-17α-methyl-17β-hydroxy-5α-androstane. This derivative (Methyldiazirinol, D-150, Lederle Laboratories) was reported [123] to possess 300% of the anabolic and 20% of the androgenic potency of 17α-methyltestosterone when administered orally.

DISSOCIATION OF ANABOLIC AND ANDROGENIC EFFECTS

Is it possible to dissociate completely anabolic from androgenic effects?

---

*I am very grateful to Dr. E. Farkas for the endocrine data (personal communication).

This question has not been answered so far. A pure anabolic steroid devoid of androgenic properties has not yet been found.

As we have seen, chemical modification of the natural androgen, testosterone, has yielded a series of compounds some of which exhibit a favorable anabolic–androgenic ratio. Among the compounds which possess a favorable anabolic–androgenic ratio those compounds appear to be most useful which achieve the favorable anabolic–androgenic ratio by having a decreased androgenic activity and an increased anabolic activity compared to those of the natural androgen molecule, testosterone, which is also used as the standard.

In spite of the fact that certain chemical modifications were highly successful in bringing about a favorable anabolic–androgenic ratio, it is not possible as yet to make generalizations about what kind of chemical modification will enhance the anabolic activity with a simultaneous repression of androgenic properties.

CONCLUSION

Is there, in general, a relationship between structural formula and biological action? Ariens pointed out [2] that there exists a relationship between structure and action, because the structure of the drug molecule determines the physicochemical properties of the drug molecule and in turn the physicochemical interaction between the molecules of the drug and the molecules of the biological object is responsible for the biological effect of a drug. "If a lack of such a relationship is observed, it only is an apparent lack. This may be the consequence of the fact that the structural formulas used represent only in a very poor way the physicochemical properties of the drug."

We have to gain more knowledge to be able to answer this question.

CHAPTER 5

TABLES

ORDER OF TABLES

CONTENTS OF TABLES

TABLES WITH ROMAN NUMERALS

Table I (Series E) is comprised of compounds with unusual structural features.

Table II (Series S) is comprised of testosterone derivatives.

Table III (Series D) is comprised of 4,5α-dihydrotestosterone derivatives.

Table IV (Series A) is comprised of 17β-hydroxy-5α-androstane derivatives containing a ring A unsaturation, excluding testosterone derivatives.

Table V (Series N) is comprised of 19-Nortestosterone derivatives.

Compiled in these tables are, in order, the serial number, the chemical structure of the steroidal compound, the ventral prostate (V.P.), seminal vesicles (S.V.), and levator ani (L.A.) indices, the standard of biological evaluation, the literature reference number, and the route of administration in the biological evaluation. The standard of biological evaluation, usually testosterone, testosterone propionate, or 17α-methyltestosterone is expressed as 100% activity. The route of administration is subcutaneous, unless noted as oral. Unless otherwise noted, the anabolic activity is expressed as the percentage of the activity of the compound compared to 100% activity of testosterone (Testost) or testosterone propionate (Test prop) (in most cases subcutaneous administration of the drug) or 17α-methyltestosterone (17α-MT) (in most cases oral administration of the drug) in increasing the weight of the levator ani muscle of the rat. Unless otherwise noted, the androgenic activity is expressed as the percentage of the activity of the compound compared to 100% activity of testosterone (Testost) or testosterone propionate (Test prop) (in most cases subcutaneous administration of the drug) or 17α-methyltestosterone (17α-MT) (in most cases oral administration of the drug) in increasing the weight of the seminal vesicles or ventral prostate of the rat.

TABLES WITH ROMAN AND ARABIC NUMBERS

Tables II.1–II.6 are comprised of testosterone derivatives.

Tables III.1–III.6 are comprised of 4,5α-dihydrotestosterone derivatives.

Tables IV.1–IV.7 are comprised of 17β-hydroxy-5α-androstane derivatives, excluding testosterone derivatives.

Tables V.1–V.6 are comprised of 19-nortestosterone derivatives.

The chemical structure of the parent compound is given on the top of these tables. Compiled in these tables are, in order, the chemical substitution, the serial number, the percentage of average androgenic activity, the percentage of average anabolic activity, the average anabolic/androgenic ratio (quotient) Q, the standard of biological evaluation, and the route of administration (SC = subcutaneous, O = oral). Only significant anabolic/androgenic quotients are given. The anabolic/androgenic quotient is expressed as the anabolic activity of the compound compared to testosterone (T), testosterone propionate (TP), or 17α-methyltestosterone (MT) (unless otherwise noted) as measured by the percent increase in weight of the levator ani muscle in the castrated rat divided by the androgenic activity of the compound compared to testosterone, testosterone propionate, or 17α-methyltestosterone as measured by the percent increase in weight of the seminal vesicles or ventral prostate of castrated rat. (17E+N T = 17-ethyl+nortestosterone.)

Table I

Structure	Name	V.P.	S.V.	L.A.	Standard	Ref.	Adm.
E-1	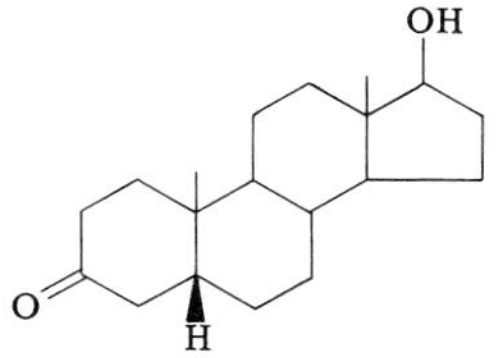 17β-Hydroxy-5β-androstan-3-one	32	11	25	Testost	80	
E-2	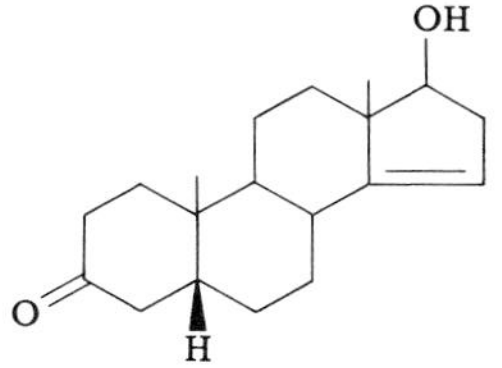 17β-Hydroxy-5β-androst-14-en-3-one	45	17	0	Testost	80	
E-3	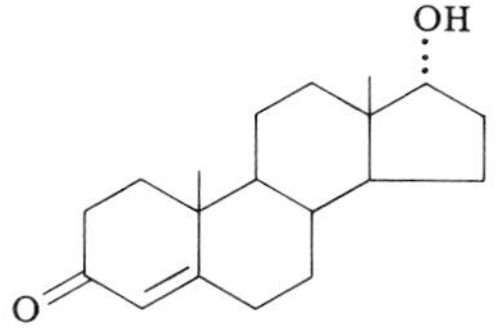 17α-Hydroxy-androst-4-en-3-one (cis-testosterone)	8 5 3.75	3 5 3.75	3 5 —	Testost Testost Testost	82 107 165	
E-4	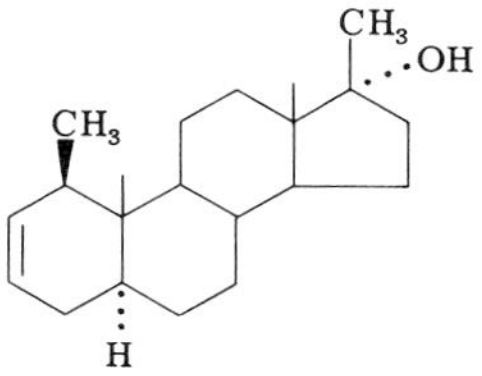 1β,17β-Dimethyl-17α-hydroxy-5α-androst-2-ene	—	<1.5	<2.5	Test prop	100	

Structure of E-1

Structure of E-2

Structure of E-3

Structure of E-4

97

Table I (continued)

Structure	Name	V.P.	S.V.	L.A.	Standard	Ref.	Adm.
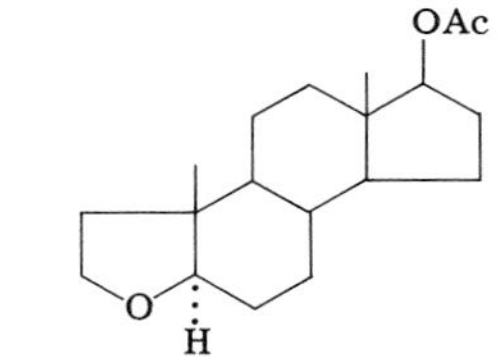	3-Oxa-5α-A-nor-androstane-17β-ol acetate	0	0	0	Test prop	83	
	2-Acetyl-7-oxo-1, 2, 3, 4, 4a, 4b, 5, 6, 7, 9, 10, 10a-dodecahydro-phenanthrene	←——— Increases all ———→ Increase in weight of chick comb (inunction) 3-4% of testost			No standard	20 21	
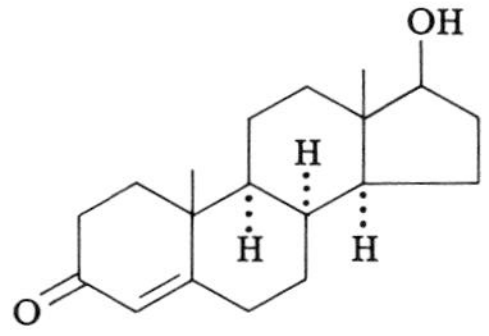	8-Isotestosterone	40% of Testost in chick comb test			(andro-genicity)	16	
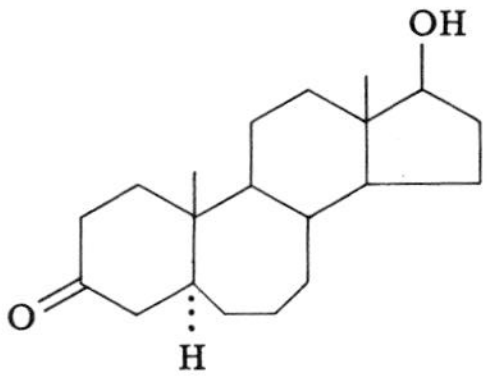	B-Homo-4,5α-di-hydrotestosterone	100	100	200	Testost	111	

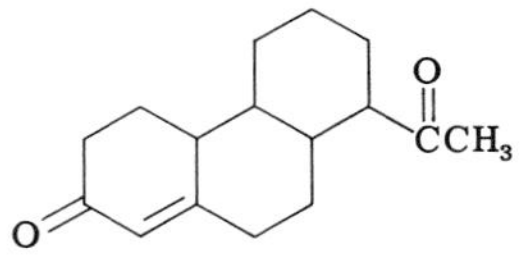

E-5

E-6

E-7

E-8

98

E-9	D-Homo-4,5α-di-hydrotestosterone	120 100 80	116 100 80	— — —	Testost Testost Testost	112 165 194
E-10	A-Homo-4,5α-di-hydrotestosterone	< 5	< 5	—	Testost	113
E-11	DL-19-Nor-D-homo-testosterone	85	—	130	Test prop	114
E-12	17β-Hydroxy-9β,10α-estr-4-en-3-one	1	1	1	Test prop	116

Table I (continued)

Structure	Name	V.P.	S.V.	L.A.	Standard	Ref.	Adm.
E-13	16,17-Seco-14β-androst-4-en-3-one-17-oic acid	—	1–3 Chick comb	—	Testost	117	
E-14	16,17-Seco-5β,14β-androst-1-en-3-one-17-oic acid	—	1–3 Chick comb	—	Testost	117	
E-15	3β-Hydroxyandrost-5-en-17-ylidene-N-pyrrolidinium p-toluenesulfonate	Weak	Weak	Weak	Test prop	118	
E-16	6α-Fluoro-17β-acetoxy-9β,10α-androst-4-en-3-one	←——— Favorable ratio ———→ No details as to the method used, but see nonclassical evaluation				119 201	

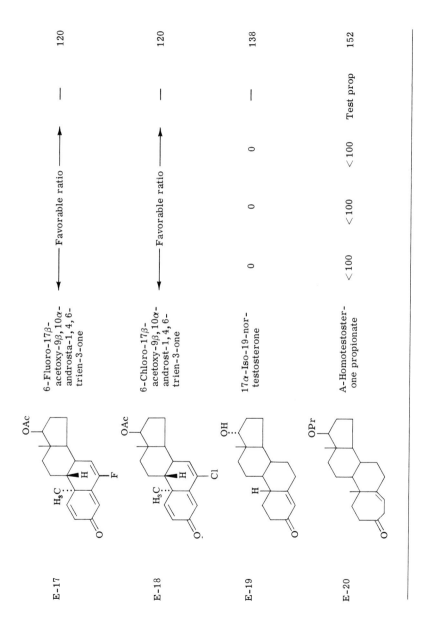

E-17   6-Fluoro-17β-acetoxy-9β,10α-androsta-1,4,6-trien-3-one   ←— Favorable ratio —→   —   120

E-18   6-Chloro-17β-acetoxy-9β,10α-androst-1,4,6-trien-3-one   ←— Favorable ratio —→   —   120

E-19   17α-Iso-19-nor-testosterone   0   0   0   —   138

E-20   A-Homotestosterone propionate <100 <100 <100 Test prop   152

Table I (*continued*)

Structure	Name	V. P.	S. V.	L. A.	Standard	Ref.	Adm.
E-21	17α-Methyl-17β-hydroxy-2-oxa-5α-androstan-3-one	24	24	322	17α-MT	52	Oral
		—	—	630	17α-MT	101	Oral
E-22	17α-Methyl-17β-hydroxy-2-oxaandrost-4-en-3-one	—	20	100	17α-MT	51	Oral
E-23	D-Homo-18-nor-17a-oxa-17,3-dioxo-5α-androstane	<5	<5	<5	Test prop	107	
E-24	17aβ-Methyl-17aα-hydroxy-D-homoandrost-4-en-3-one	0	0	0	Test prop	157	
		0	0	—	Testost	165	

E-25	17α-Ethyl-D-homo-nortestosterone	—	—	130	Testost	107
E-26	3,3,17,17-Tetra-fluoro-5α-andros-tane	0	0	0	Testost	107
E-27	5α-Androstano-[3,2-d]pyrimidine-[17,16-c]pyrazole	0	0	0	Testost	137
E-28	Δ³-A-Norandrosten-2-one-3β,17β-diol	0	0	—	Testost	158
	Diacetate	0	0	—	Testost	158

103

Table I (*continued*)

	Structure	Name	V. P.	S. V.	L. A.	Standard	Ref.	Adm.
E-29	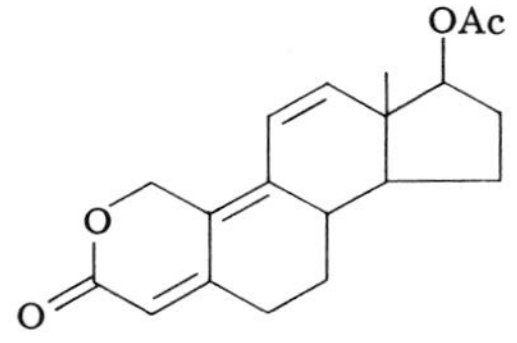	2-Oxa-17β-acetoxy-estra-4, 9, 11-trien-3-one	500	500	500	Nortestost acetate	159	
E-30	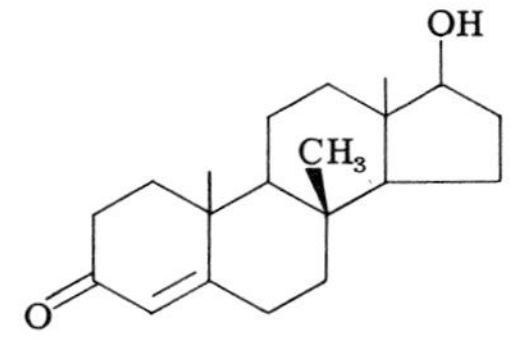	8β-Methyltestos-terone	< 1.5	< 4.5	< 3	Testost	160	
E-31	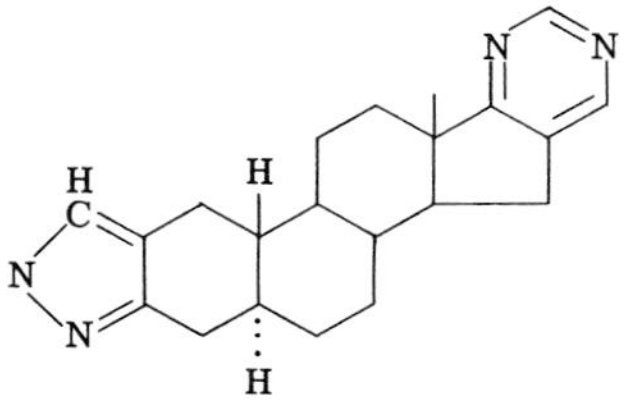	5α-Estrano[3, 2-c]-pyrazole[17, 16-d] pyrimidine	0	0	0	Testost	137	
E-32	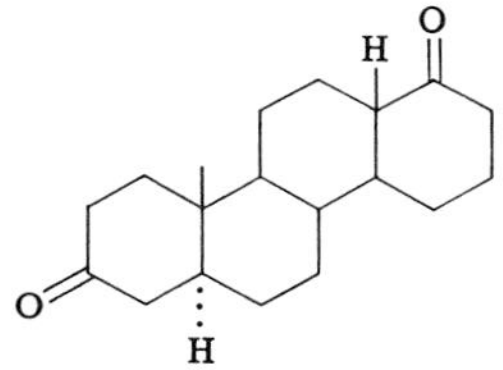	18-Nor-D-homo-androstane-3, 17a-dione	10 20	10 20	— —	Testost Testost	162 165	

104

	Compound				Reference	Ref.
E-33	17-Epitestosterone	0	0	0	Testost	164
		0		—	Testost	165
E-34	3β,17α-Dihydroxy-5α-androstane	1.5	1.5	—	D-Homo-testost	165
E-35	3α-Hydroxy-17a-keto-D-homo-5α-androstane (D-homoandrosterone)	15−17	20	—	D-Homo-testost	165
			20	—	Testost	194
E-36	3β-Hydroxy-17a-keto-D-homo-5α-androstane(D-homoepiandrosterone)	10	13	—	D-Homo-testost	165
		10		—	Testost	194

105

Table I (*continued*)

	Structure	Name	V. P.	S. V.	L. A.	Standard	Ref.	Adm.
E-37		D-Homo-3β,17aβ-dihydroxy-5α-androstane	12.5 3	12.5 3	— —	D-Homo-testost Testost	165 194	
E-38		D-Homo-3α,17aα-dihydroxy-5α-androstane	4	4	—	D-Homo-testost	165	
E-39		D-Homo-androst-4-en-3,17a-dione	50—66	50—66	—	D-Homo-testost	165	
E-40		D-Homotestoster-one	50	50	—	Testost	165	

E-41	D-Homo-*cis*-testosterone	5	5	—	Testost	165
E-42	D-Homo-17α-methyltestosterone	100	100	—	Testost	165
E-43	D-Homo-5α-androstan-3, 17a-dione	6	6	—	Testost	194
E-44	D-Homo-5α-androstan-3, 17-dione	2	2	—	Testost	194

Table I (*continued*)

	Structure	Name	V.P.	S.V.	L.A.	Standard	Ref.	Adm.
E-45		D-Homo-5α-androstan-3β,17aα-diol	9	9	—	Testost	194	
E-46		6-(4'-Ketocyclohexyl)-Δ^{1,9}-octalone	<1	<1	—	Test prop	195	
E-47		D-Homo-19-nor-androst-4-en-3,17a-dione	15	—	15	Test prop	114	
E-48		3-Methoxy-D-homo-estra-2,5(10)-dien-17aβ-ol	77.5	—	125	Test prop	114	

E-49	8α,10α-Testos-terone	0	0	0	Testost	199
E-50	9β,10α-Testos-terone Retrotestosterone	0	0	0	Testost	198
E-51	13α-Androst-5-en-3β-ol-17-one acetate	0	0	0	Testost	202
E-52	13α-Androst-4-en-3,17-dione	0	0	0	Testost	202

109

Table I (continued)

Structure	Name	V.P.	S.V.	L.A.	Standard	Ref.	Adm.
E-53	13α-Androst-4-en-17α-ol-3-one	0	0	0	Testost	202	
E-54	13α-Androst-4-en-17β-ol-3-one	0	0	0	Testost	202	
E-55	A-Homotestosterone acetate	← No substitution of testosterone deficit → after castration was possible				294	
E-56	A-Nor-17β-acetoxy-estra-4,9,11-trien-3-one	500	500	500	Nortestost acetate	159	

E-57	8β-Formyltestoster-one	<1.5	<4.5	<14	Testost	160
E-58	4α,5α-Ethylene-17β-acetoxy-androstan-3-one	<25	<25	<25	Testost	378
E-59	6α,7α-Vinylene-17β-acetoxyandrost-4-en-3-one	<20	<20	<20	Testost	378
E-60	6β,7β-Vinylene-17β-acetoxyandrost-4-en-3-one	<20	<20	<20	Testost	378

Table II

Number	Structure	Name	V.P.	S.V.	L.A.	Standard	Ref.	Adm.
S-1		Testosterone	—	35	26	Test prop	36	
			—	100	60	17α-MT	45	Oral
			28	50	36	17α-MT	30	Oral
			—	145	100	17α-MT	36	Oral
S-2		17α-Methyl-testosterone	103	94	115	Testost	30	
			168	—	159	Test prop	81	
			356	200	280	Testost	30	Oral
			130	130	150	Testost	64	
S-3		Testosterone propionate	165	168	151	Testost	30,82	
			—	285	385	Testost	46	
			99	93	96	Testost	115	
			—	275	385	Testost	45	
S-4		17α-Ethyl-testosterone	5	5	10	Test prop	31	
			2.8	—	5	Test prop	91	
			—	—	30	17α-MT	107	

	Compound						
S-5		11β-Hydroxy-testosterone	35	33	40	Testost	30
S-6		4-Chloro-testosterone propionate	42 50	42 50	82.5 100	Test prop Testost	32 78
S-7		4-Chloro-11β-hydroxy-testosterone acetate	38 14	32 14	55 68	Test prop Test prop	32 107
S-8		4-Bromo-testosterone	20	31	26	Test prop	32

113

Table II (continued)

Number	Structure	Name	V.P.	S.V.	L.A.	Standard	Ref.	Adm.
S-9		4-Fluoro-testosterone acetate	13 —	32 —	24 45	Test prop Test prop	32 107	
S-10		4-Hydroxy-testosterone acetate	28 —	24 —	52 80	Test prop Test prop	32 107	
S-11		4-Methyl-testosterone	10 40 40	10 40 40	30 120 120	Test prop Testost Testost	33 78 27	
S-12		6α-Methyl-testosterone	90 <100 23	90 <100 23	460 <100 42	Testost Testost Testost	34 25 107	

No.	Compound				Standard		Route
S-13	4-Ethyl-testosterone	0	0	0	Test prop	33	Test prop · 33
S-14	4-Allyl-testosterone	0	0	0	Test prop	33	Test prop · 33
S-15	Δ^1-Dehydro-17α-methyl-testosterone	0.8	0.8	6.0	Test prop	84	
		—	73	182	17α-MT	76	Oral
		55	50	125	17α-MT	30	Oral
		60	60	210	17α-MT	9	Oral
		6	12	30	17α-MT	30	Oral
		—	45	89	17α-MT	42	Oral
S-16	4-Hydroxy-17α-methyl-testosterone	61	42	134	17α-MT	30	Oral
		—	56	230	17α-MT	76	Oral
		50	50	330	17α-MT	26	Oral
		—	52	165	17α-MT	127	[Also see ref. 105 (Gland)]

Table II *(continued)*

Number	Structure	Name	V.P.	S.V.	L.A.	Standard	Ref.	Adm.
S-17	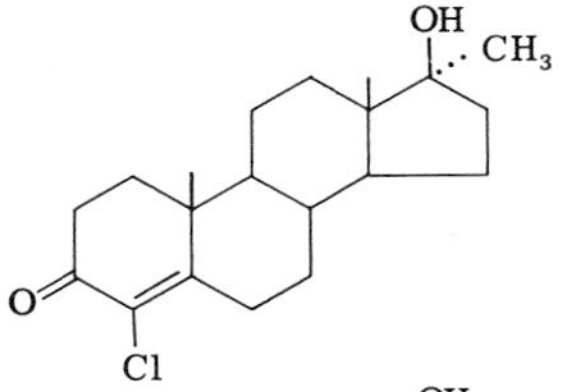	4-Chloro-17α-methyl-testosterone	27 15 12	26 10 —	46 50 32	17α-MT 17α-MT 17α-MT	30 30 44	Oral Oral
S-18	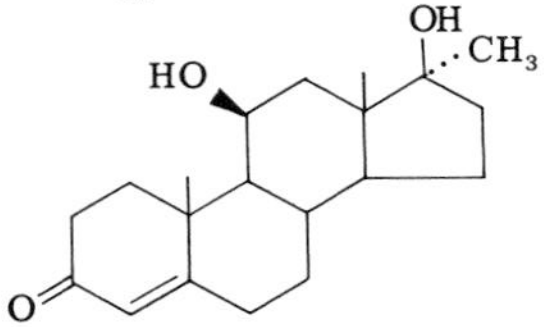	11β-Hydroxy-17α-methyl-testosterone	33 —	35 90	40 300	17α-MT 17α-MT	30 37	Oral
S-19	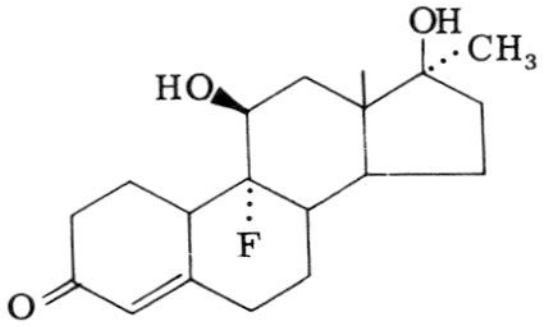	9α-Fluoro-11β-hydroxy-17α-methyl-testosterone	75 — 120 46-48	188 950 750 64-93	89 2000 1745 72-175	17α-MT 17α-MT 17α-MT Testost	30 37 30 61	[Also see ref. 105 (Gland)] Oral Oral
S-20	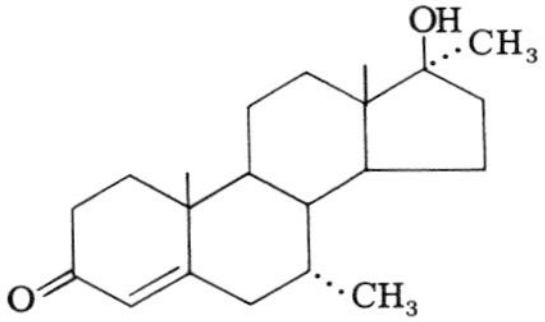	7α,17α-Dimethyl-testosterone	170 224 —	210 660 300	146 1140 575*	Testost Testost 17α-MT	35 35 72	[Also see ref. 105 (Gland)] Oral Oral

*Monkey anabolic evaluation

116

S-21	7α-Methyl-testosterone	115	133	127	Testost	35	Oral
		68	36	110	Testost	35	
S-22	Δ⁴-Androsten-3,17-dione	131	63	55	Testost	35	Oral
		110	7	53	Testost	35	
		39	17	22	Testost	61	
		32	30	32	Testost	62	
S-23	7α-Methyl-Δ⁴-androsten-3,17-dione	60	30	77	Testost	35	Oral
		62	78	225	Testost	35	
S-24	17α-Methyl-11-keto-testosterone	—	65	140	17α-MT	37	Oral

117

Table II (*continued*)

Number	Structure	Name	V. P.	S. V.	L. A.	Standard	Ref.	Adm.
S-25	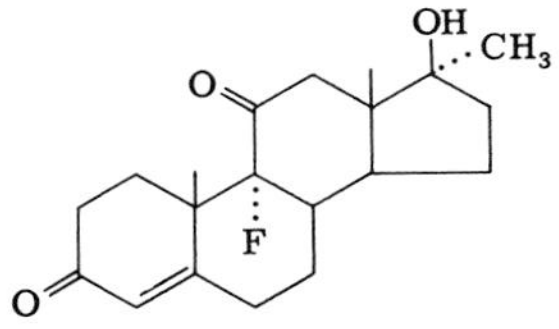	9α-Fluoro-11-keto-17α-methyl-testosterone	—	850	2200	17α-MT	37	Oral
S-26	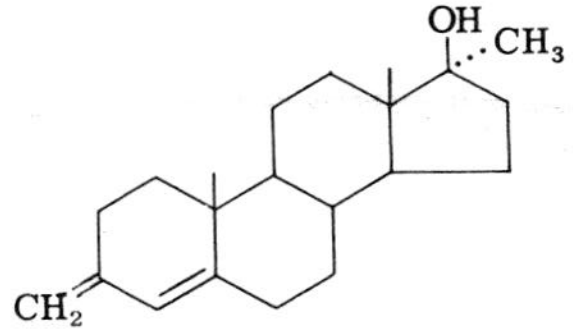	3-Methylene-17α-methyl-androst-4-en-17β-ol	—	28	53	17α-MT	48	
S-27	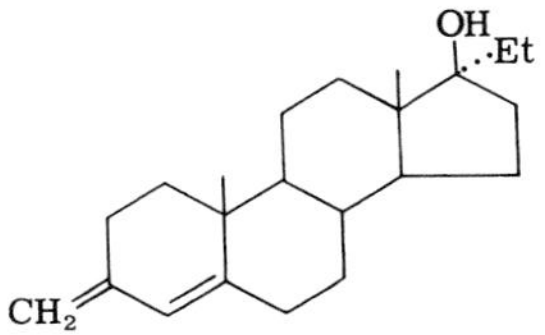	3-Methylene-17α-ethyl-androst-4-en-17β-ol	—	0.8	4	17α-MT	48	
S-28	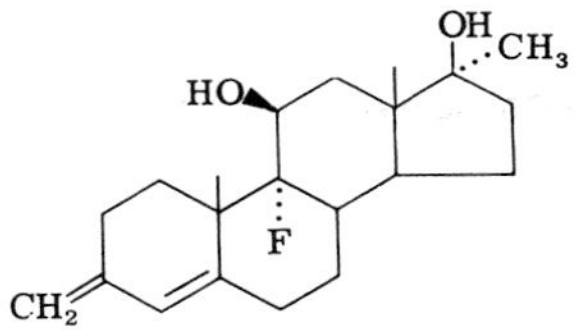	9α-Fluoro-3-methylene-17α-methyl-androst-4-en-11β,17β-diol	—	32	30	17α-MT	48	

118

Compound	Structure				Standard		
S-29	6α-Chloro-testosterone-17-acetate	190	120	270	Testost	56	
		80	80	300	Testost	53	
S-30	Testosterone 17-dichloroacetate	350	350	350	Testost	56	
		1030	769	786	Testost	82	
S-31	6α-Fluoro-testosterone	200	170	190	Testost	56	
		50	50	100	Testost	78	
		123	128	206	Testost	82	
		—	—	230	17α-MT	18	Oral
S-32	2-Methylene-17α-methyl-17β-hydroxyandrost-4-en-3-one	<40	<50	100	Testost	56	
		15	15	50	17α-MT	98	Oral

119

Table II *(continued)*

Number	Structure	Name	V.P.	S.V.	L.A.	Standard	Ref.	Adm.
S-33		2α-Fluoro-testosterone	30 20	20 20	30 50	Testost Testost	60 28	
S-34		2α-Fluoro-17α-ethynyltestosterone	10	20	10	Testost	60	
S-35		2-Hydroxy-methylene-17α-ethynyltestosterone	<10	<10	<10	Testost	60	
S-36		2-Hydroxy-methylene-17α-propynyl-testosterone	<10	<10	<10	Testost	60	

	Compound					
S-37	1α- Cyano - androst - 4 - en - 3,17 - dione	<10	<10	<10	Testost	60
S-38	1 - β - Thioacetyl- androst - 4 - en - 3,17 - dione	<10	<10	<10	Testost	60
S-39	6β - Nitro- testosterone acetate	<10	<10	<10	Testost	60
S-40	6α - Nitro- testosterone	<10	<10	<10	Testost	60

121

Table II (continued)

Number	Structure	Name	V.P.	S.V.	L.A.	Standard	Ref.	Adm.
S-41		2α-Dimethylamino-testosterone acetate	<10	<10	<10	Testost	60	
S-42		2α-Dimethylamino-testosterone acetate methiodide	<10	<10	<10	Testost	60	
S-43		Androst-4-ene-3,11,17-trione [adrenosterone]	7 48	8 48	10 70	Testost Testost	61 62	
S-44		11β-Hydroxy-androst-4-en-3,17-dione	<12 <10	<12 <10	<12 20	Testost Testost	61 62	

No.	Structure	Name				Reference		
S-45		17α-Ethynyl-testosterone	10 0	8 —	4 1	Testost Test prop	64 91	Oral
S-46		1α-Methylthio-17α-methyltestosterone	—	82	194	17α-MT	76	Oral
S-47		1α-Ethylthio-17α-methyltestosterone	—	35	66	17α-MT	76	Oral
S-48		1α-Acetylthio-17α-methyltestosterone	—	33	101	17α-MT	76	Oral

Table II (continued)

Number	Structure	Name	V.P.	S.V.	L.A.	Standard	Ref.	Adm.
S-49		4 - Mercapto - 17α - methyl- testosterone	—	17	18	17α-MT	76	Oral
S-50		4 - Methylthio - 17α - methyl- testosterone	—	31	46	17α-MT	76	Oral
S-51		4 - Ethylthio - 17α - methyl- testosterone	—	11	24	17α-MT	76	Oral
S-52		4 - Acetylthio - 17α - methyl- testosterone	—	22	47	17α-MT	76	Oral

No.	Structure	Name		Androgenic	Anabolic	Standard	Ref.	Route
S-53		7α-Mercapto-17α-methyl-testosterone	—	34	144	17α-MT	76	Oral
S-54		7α-Methylthio-17α-methyl-testosterone	—	7	26	17α-MT	76	Oral
S-55		7α-Ethylthio-17α-methyl-testosterone	—	20	90	17α-MT	76	Oral
S-56		7α-Acetylthio-17α-methyl-testosterone	—	27	62	17α-MT	76	Oral

125

Table II (continued)

Number	Structure	Name	V.P.	S.V.	L.A.	Standard	Ref.	Adm.
S–57	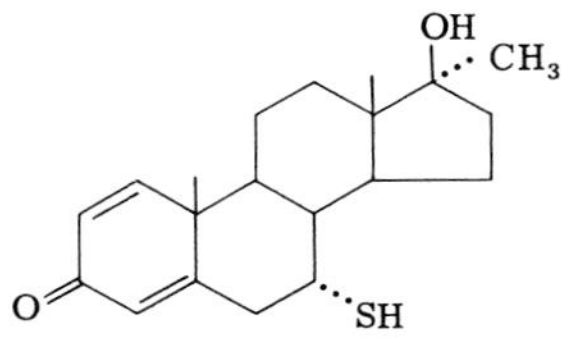	1-Dehydro-7α-mercapto-17α-methyltestosterone	—	59	62	17α-MT	76	Oral
S–58	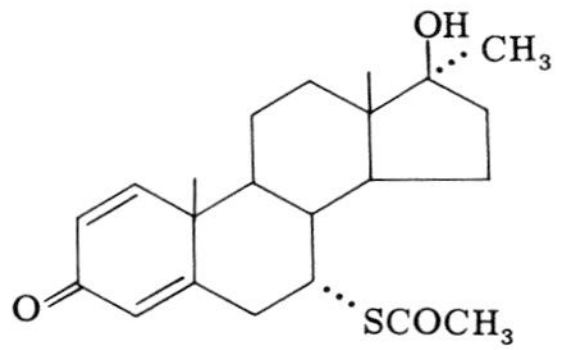	1-Dehydro-7α-acetylthio-17α-methyltestosterone	—	43	75	17α-MT	76	Oral
S–59	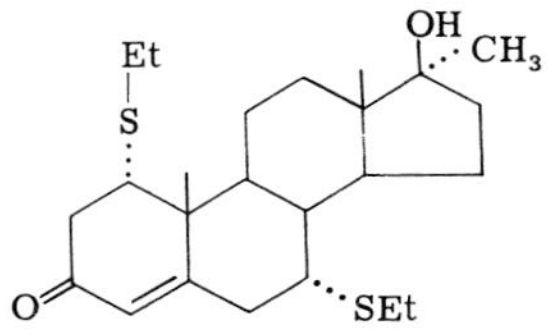	1α,7α-Bis (ethylthio) - 17α- methyl-testosterone	—	22	122	17α-MT	76	Oral
S–60	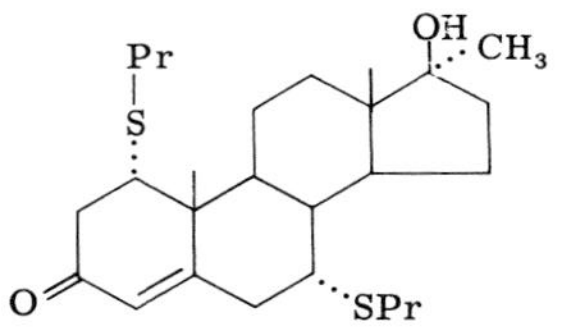	1α,7α-Bis (propylthio) - 17α- methyl-testosterone	—	8	40	17α-MT	76	Oral

126

	Structure	Name						
S-61		1α,7α-Bis(acetyl-thio)-17α-methyl-testosterone "Emdabol"	—	61	456	17α-MT	76	Oral
S-62		6β-Fluoro-testosterone	25	25	25	Testost	78	
			30	30	30	Testost	107	
S-63		Δ14-Testosterone	85	60	48	Testost	80	Oral
			108	190	125	Testost	80	
S-64		6-Chloro-17β-acetoxyandrosta-4,6-dien-3-one	14	42	119	Testost	82	

127

Table II (continued)

Number	Structure	Name	V. P.	S. V.	L. A.	Standard	Ref.	Adm.
S-65	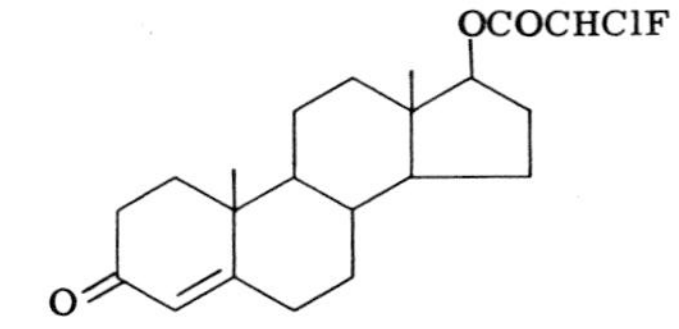 OCOCHClF	Testosterone 17-fluorochloro-acetate	233	361	312	Testost	82	
S-66	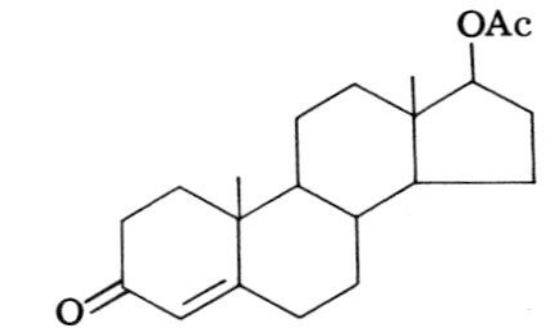 OAc	Testosterone 17-acetate	97	93 100–200	74 100–200	Testost Test prop	82 100	
S-67	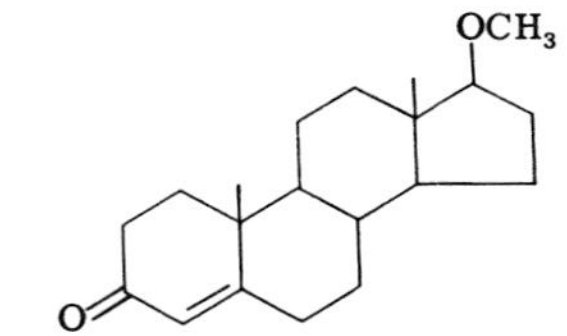 OCH₃	Testosterone 17-methyl ether	34 <100	23 <100	23 <100	Testost Test prop	82 153	
S-68	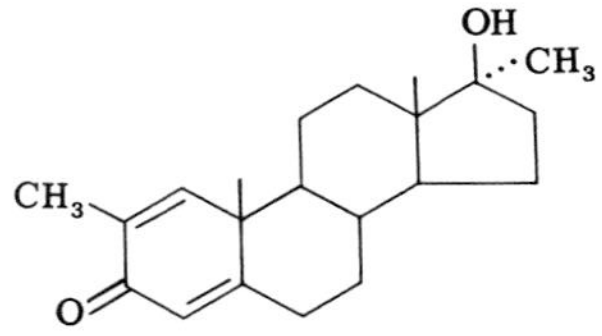 OH ···CH₃ CH₃	2,17α-Dimethyl-17β-hydroxy-androsta-1,4-dien-3-one	22 40	27 40	49 42	Testost Testost	82 107	

128

No.	Compound				Standard		
S-69	2,17α - Dimethyl 17β - hydroxy - androsta - 1,4,6 - trien - 3 - one	130	115	300	Testost	82	
S-70	17β - Hydroxy - androsta - 1,4 - dien - 3 - one	0.4	0.8	4.3	Test prop	84	
		9	12	25	17α-MT	84	
		54	52	100	Testost	107	Oral
S-71	6β - Chloro - 17β - acetoxyandrost - 4 - en - 3 - one	20	20	20	Testost	53	
S-72	1α - Methyltestosterone acetate	—	15	130	Test prop	100	

129

Table II (continued)

Number	Structure	Name	V. P.	S. V.	L. A.	Standard	Ref.	Adm.
S-73		1β-Methyltestos-terone acetate	—	6	50–100	Test prop	100	
S-74		1α-Chloromethyl-testosterone acetate	—	<5	60?	Test prop	100	
S-75		1α-Bromomethyl-testosterone acetate	—	<1.5	10–20	Test prop	100	
S-76		1α-Iodomethyl-testosterone acetate	—	1.5	30–40	Test prop	100	

S-77		1,1 - Dimethyl-testosterone acetate	—	< 1.5	< 2.5	Test prop	100
S-78		1,2α - Methylene-testosterone acetate	—	5	50	Test prop	100
S-79		4 - Chloro - 1α - methyltestosterone acetate	—	10	50	Test prop	100
S-80		1α,7α- Dimethyl-testosterone acetate	—	40	300-400	Test prop	100

Table II (continued)

Number	Structure	Name	V. P.	S. V.	L. A.	Standard	Ref.	Adm.
S-81		$1\alpha,7\beta$ - Dimethyl-testosterone acetate	—	<4.5	<7.5	Test prop	100	
S-82		1α - Methyl - 11β-hydroxytestosterone acetate	—	<4.5	10–20	Test prop	100	
S-83		$1,2\alpha$ - Methylene - 11β-hydroxytestosterone acetate	—	<4.5	50	Test prop	100	
S-84		1α- Methyl - 11 - oxo-testosterone acetate	—	2	20–30	Test prop	100	

					Test prop	100

S-85 — 1-Methyl-Δ^1-testosterone acetate — | <4.5 | <7.5 | Test prop | 100

S-86 — 1α-Methyl-Δ^6-testosterone acetate — | <4.5 | <7.5 | Test prop | 100

S-87 — 1-Methyl-$\Delta^{1,6}$-testosterone acetate — | <4.5 | <7.5 | Test prop | 100

S-88 — 1α-Methyl-11-oxo-Δ^6-testosterone acetate — | <4.5 | <7.5 | Test prop | 100

133

Table II (continued)

Number	Structure	Name	V.P.	S.V.	L.A.	Standard	Ref.	Adm.
S-89		1,2α - Methylene - Δ^6 - testosterone acetate	—	1.5	10	Test prop	100	
S-90		4 - Chloro - 1α - methyl - Δ^6 - testosterone acetate	—	<4.5	<7.5	Test prop	100	
S-91		6 - Chloro - 1,2α - methylene - Δ^6 testosterone acetate	—	<1.5	6	Test prop	100	
S-92		1α,17α-Dimethyl-testosterone	—	2	20	Test prop	100	

	Compound					
S-93	1,17α-Dimethyl-Δ^1-testosterone acetate	—	<1.5	<2.5	Test prop	100
S-94	1α,7α,17α-Trimethyl-testosterone acetate	—	40	500	Test prop	100
S-95	6β-Methyl-testosterone	<100	<100	<100	Testost	25
S-96	6α,17α-Dimethyl-testosterone	<100	65	<100	Testost	25
			65	58	Testost	107

135

Table II (continued)

Number	Structure	Name	V.P.	S.V.	L.A.	Standard	Ref.	Adm.
S-97		6β,17α-Dimethyl-testosterone	<100	<100	<100	Testost	25	
S-98		17α - Ethynyl - 17β-acetoxy - 11β-hydroxyandrost - 4 - en - 3 - one	0	0	0	Testost	17	
S-99		17α - Ethyl - 11β - hydroxy - testosterone acetate	0	0	0	Testost	17	
S-100		16,16 - Difluoro - 17β - hydroxyandrost - 4 - en - 3 - one	—	2.5	0	Test prop	104	

	Structure	Name			Standard	Value	Route
S-101		17α- Cyclopropyl-testosterone	Weak androgen	—	17α-MT	392	Oral
S-102		4 - Chloro -17α-methyl- 11β,17β - dihydroxyandrost- 4 - en - 3 - one	⟨More effective⟩ 6	— >30	Δ¹ - MT	109 309	Oral Oral
S-103		4 - Chloro -17α-methyl - 11β,17β - dihydroxyandrosta - 1,4 - dien - 3 - one	⟨Less effective⟩ 0	—	Δ¹ - MT	109	Oral
S-104		11β - Methyl-testosterone	0	0	Test prop	161	

Table II (continued)

Number	Structure	Name	V. P.	S. V.	L. A.	Standard	Ref.	Adm.
S-105	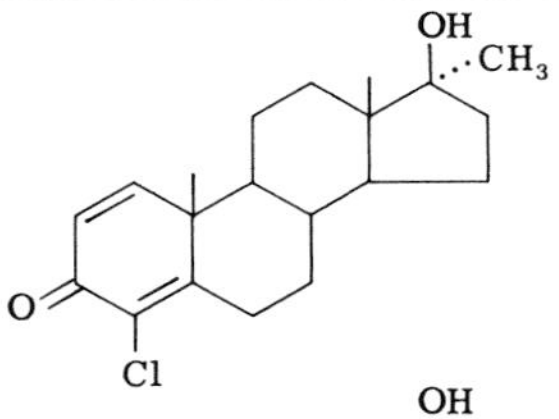	4 - Chloro - 17α - Methyl - 17β - hydroxyandrosta - 1,4 - dien - 3 - one	—	—	> 100	17α - MT	110 234 263	Oral
S-106	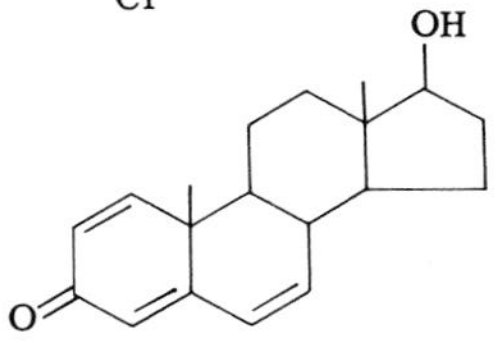	17β - Hydroxy - androst - 1,4,6 - trien - 3 - one	< 33	< 33	< 33	Testost	107	
S-107	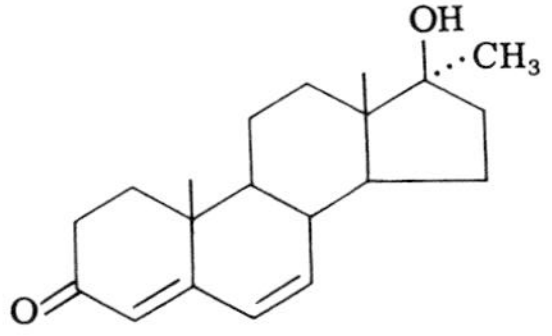	17α - Methyl - 17β - hydroxyandrosta - 4,6 - dien - 3 - one	5.1 24	— —	7.2 61	17α - MT 17α - MT	124 124	Oral
S-108	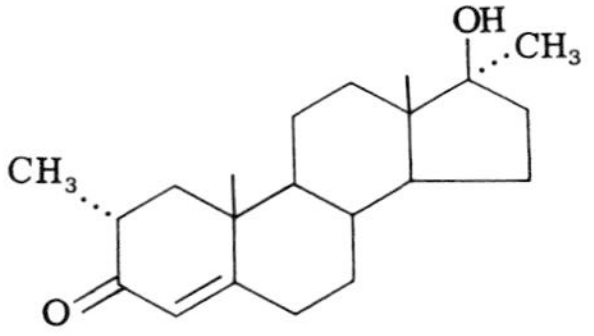	2α,17α - Dimethyl - 17β - hydroxyandrost - 4 - en - 3 - one	1 —	1 —	23 100	Testost 17α - MT	107 107	

138

		Compound					
S-109		2α-Cyano-17β-hydroxyandrost-4-en-3-one	< 5	< 5	< 5	Testost	107
S-110		17β-Acetoxy-2-chloroandrost-1,4-dien-3-one	< 5	< 5	< 5	Testost	107
S-111		$2\alpha,4$-Dichloro-17β-hydroxy-androst-4-en-3-one	Weak	Weak	Weak	Testost	107
S-112		$2\alpha,6\alpha$-Dichloro-17β-hydroxy-androst-4-en-3-one	Weak	Weak	Weak	Testost	107

139

Table II *(continued)*

Number	Structure	Name	V. P.	S. V.	L. A.	Standard	Ref.	Adm.
S-113	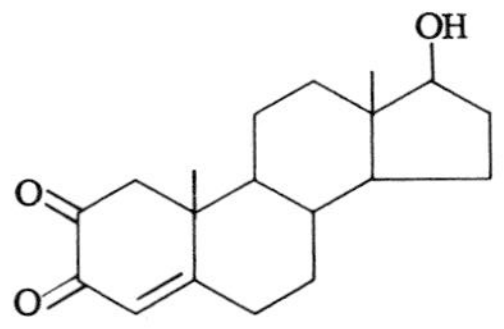	2 - Keto- testosterone	—	—	5	Test prop	107	
S-114	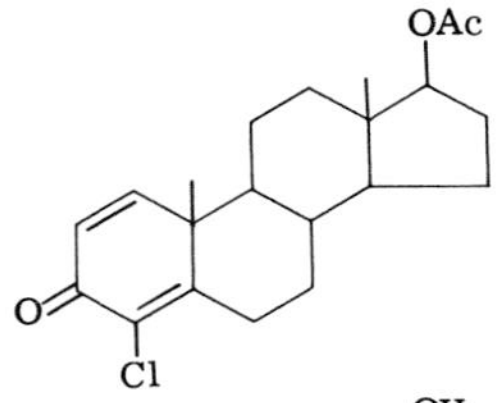	4 - Chloro - 17β - acetoxyandrost - 1,4 - dien - 3 - one	30	30	80	Test prop	107	
S-115	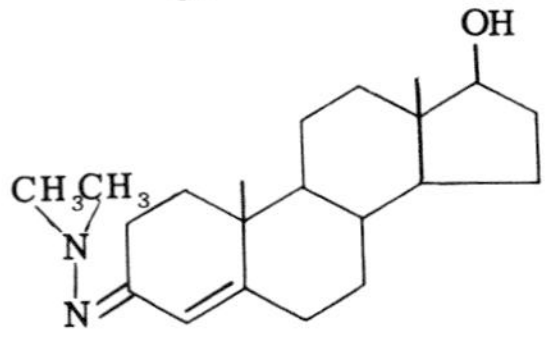	Testosterone dimethyl- hydrazone	11	2	9	Test prop	128	
S-116	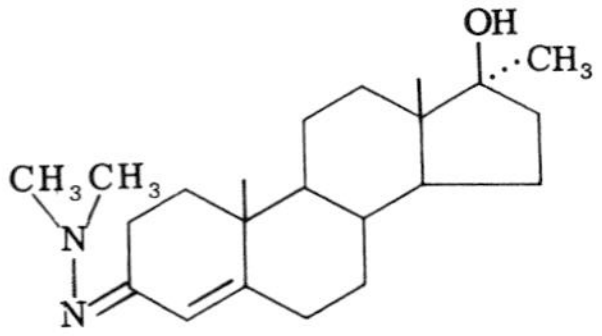	17α - Methyl- testosterone dimethyl- hydrazone	32	18	8	Test prop	128	

140

S-117	6β,17β-Diacetoxy-androst-4-en-3-one	—	< 5	Testost	107	
S-118	6α-Fluoro-17α-methyl-17β-hydroxyandrost-4-en-3-one	—	230	17α-MT	18	Oral
S-119	7α-Methylthio-testosterone acetate	140 40	90 10	17α-Ethyl-nortest	133 133	Oral
S-120	7α-Acetylthio-testosterone acetate	40	50	17α-Ethyl-nortest	133	

141

Table II (continued)

Number	Structure	Name	V. P.	S. V.	L. A.	Standard	Ref.	Adm.
S-121		7α-Mercapto-testosterone acetate	110	—	80	17α-Ethyl-nortest	133	
S-122		9α-Fluoro-11β-hydroxytestosterone acetate	—	—	35	Test prop	107	
S-123		9α-Bromo-11β-chloro-Δ¹-testosterone	—	4	34	Test prop	134	
S-124		9α-Bromo-11β-fluoro-Δ¹-testosterone propionate	—	6	77	Test prop	134	

142

S-125		9α-Chloro-11β-fluorotestosterone	—	51	84	Test prop	134
S-126		17α-Methyl-9α-bromo-11β-fluorotestosterone	—	4	17	17α-MT	134
S-127		9α,11β-Dichloro-17α-methyl-Δ¹-testosterone	—	60	300	17α-MT	134
S-128		9α-Bromo 11β-fluoro-17α-methyl-Δ¹-testosterone	—	12	23	17α-MT	134

Table II (continued)

Number	Structure	Name	V. P.	S. V.	L. A.	Standard	Ref.	Adm.
S-129	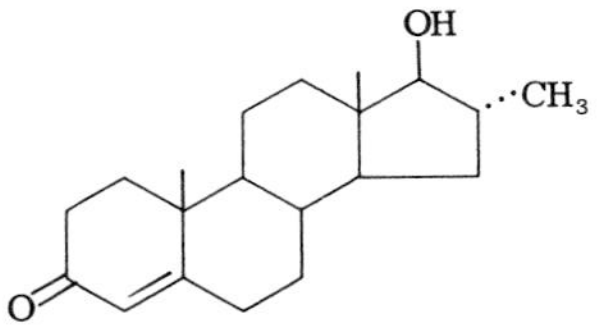	16α - Methyl - testosterone	<100	<100	<100	Testost	135	
S-130	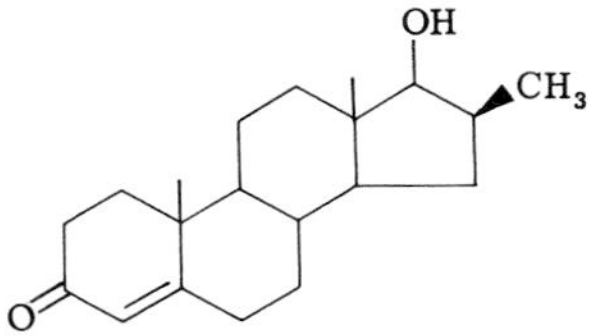	16β - Methyl - testosterone	<100	<100	<100	Testost	135	
S-131	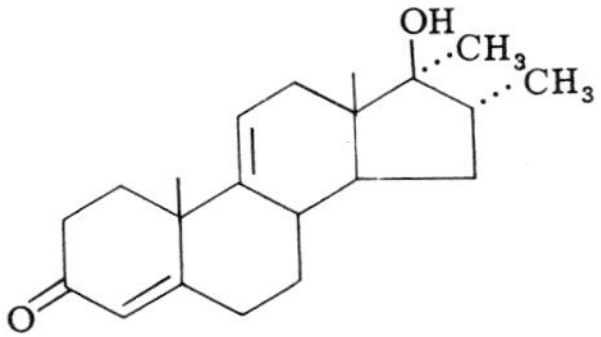	16α,17α - Dimethyl - 17β - hydroxy - androsta - 4,9(11) - dien - 3 - one	—	17	25	Fluoxy- mesterone	136	Oral
S-132	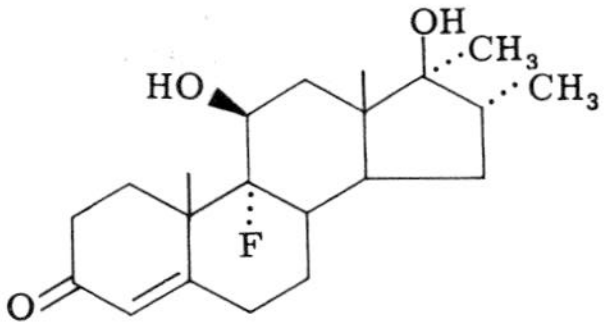	9α - Fluoro - 11β,17β - dihydroxy - 16α,17α - dimethylandrost - 4 - en - 3 - one	—	0	14	Fluoxy- mesterone	136	Oral

144

	Name				Method	Dose	Route
S-133	Testosterone 17β-(1'-ethoxy)-cyclopentyl ether	—	190	210	17α–MT	12	Oral
S-134	Testosterone 17-cyclopent-1'-enyl ether	—	130	110	17α–MT	12	Oral
S-135	Testosterone 17-(1'-ethoxy)-cyclohexyl ether	—	110	220	17α–MT	12	Oral
S-136	Testosterone 17-trimethyl-silyl ether	133	Peak at 10-30 days 209	130	Test prop Peak7-10 days	149 397	

OEt (S-133) · O (S-134) · OEt (S-135) · OSi(CH$_3$)$_3$ (S-136)

Table II (continued)

Number	Structure	Name	V. P.	S. V.	L. A.	Standard	Ref.	Adm.
S-137		2-Formyl-17α-methyl-17β-hydroxyandrosta-1,4-dien-3-one	10	10	10	Testost	155	Oral
S-138		7α,17α-Dimethyl-17β-hydroxy-androsta-1,4-dien-3-one	—	129	332	17α-MT	156	Oral
S-139		11α-Methyl-11β-hydroxytestosterone	0	0	0	Test prop	161	
S-140		17α-Propargyl-testosterone	0	0	—	—	163	

S-141	16α- Fluoro - 16β - methyl- testosterone	10	30	0	Testost	96
S-142	6α- Acetoxy - androst - 4 - en - 3,17 - dione	0*	⟨*By chick comb⟩		—	231
S-143	6β - Acetoxy - androst - 4 - en - 3,17 - dione	0*	⟨*By chick comb⟩		—	231
S-144	6β - Hydroxy - androst - 4 - en - 3,17 - dione	0	—	—	—	231

Table II (continued)

Number	Structure	Name	V. P.	S. V.	L. A.	Standard	Ref.	Adm.
S-145	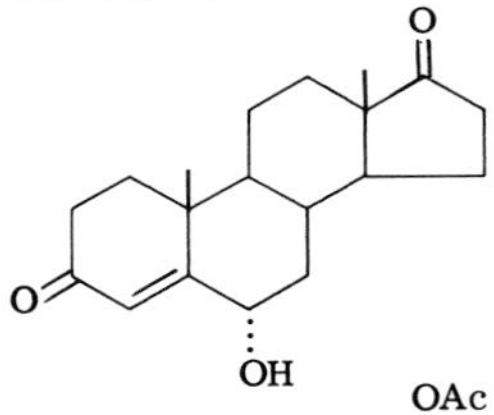	6α - Hydroxy - androst - 4 - en - 3,17 - dione	0	—	—	—	231	
S-146	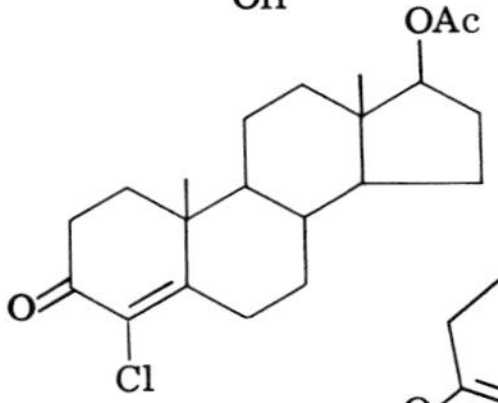	4 - Chloro - testosterone acetate	55 — — 157*	48.3 14 3 ⟨*By lachrymal gland⟩	126 68 10	Test prop Test prop Test prop Testost	32 243 379 105	
S-147	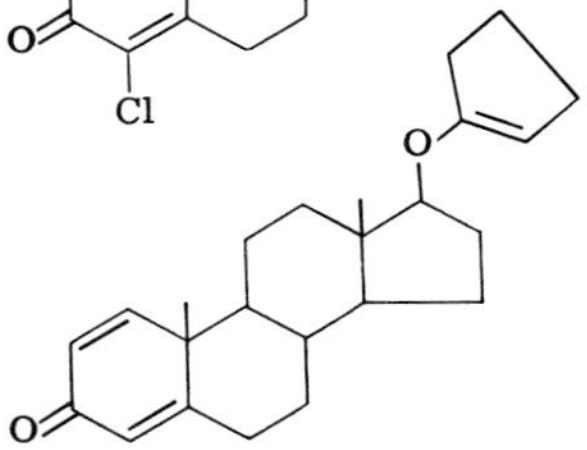	17β - Hydroxy - androst-1,4 - dien-3-one 17 - cyclopent-1'- enyl ether	—	20	60	17α - MT	12	Oral
S-148	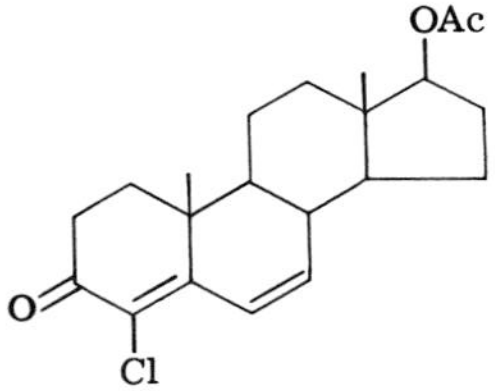	Δ⁶-Chloro - testosterone acetate	<10	0	0	Test prop	243	

148

No.	Structure	Name				Standard	Ref.	Route
S-149		4-Chloro-11-keto-17α-methyl-testosterone	—	6	730	Δ¹-17α-MT	309	Oral
S-150		1α-Methyl-testosterone	30	27	60	Testost	398	
S-151		1α-Ethyl-testosterone	7	2	2	Testost	398	
S-152		6-Chloro-17β-acetoxy-1α,2α-methylene-4,6-androstadien-3-one	⟨Antiandrogenic⟩		—		407	

149

Table II.1

Compound	Serial No.	Androgenic activity	Anabolic activity	Q	Standard	Rte
Parent	S-1	100	100	1	Definition	SC
Parent	S-1	35	26	0.7	TP	SC
Parent	S-1	50-100	36-60	0.65	MT	O
1α CH$_3$	S-150	30	60	2	T	SC
1α Et	S-151	<10	<10	—	T	SC
Δ^1	S-70	50	100	2	T	SC
Δ^i	S-70	10	25	2.5	MT	O
Δ^1, Δ^6	S-106	<33	<33	1	T	SC
Δ^1, 9α Br, 11β Cl	S-123	5	35	7	TP	SC
2α F	S-33	20	40	2	T	SC
2α CN	S-109	<2	< 2	—	T	SC
2 Oxo	S-113	—	5	—	TP	SC
2α Cl, 4 Cl	S-111	<10,	<10	—	T	SC
2α Cl, 6α Cl	S-112	—	5	—	TP	SC
3-Dimethylhydrazone	S-115	6	9	1.5	TP	SC
4 Br	S-8	25	25	1	TP	SC
4 CH$_3$	S-11	40	120	3	T	SC
4 Et	S-13	0	0	—	TP	SC
4 Allyl	S-14	0	0	—	TP	SC
6α CH$_3$	S-12	25	40	1.6	T	SC
6α F	S-31	50-200	100-200	1-2	T	SC
6α NO$_2$	S-40	<10	<10	—	T	SC
6β CH$_3$	S-95	<100	<100	1	T	SC
6β F	S-62	25	25	1	T	SC
7α CH$_3$	S-21	120	120	1	T	SC
7α CH$_3$	S-21	50	110	2	T	O
9α Cl, 11β F	S-125	50	80	1.6	TP	SC
11α CH$_3$, 11β OH	S-139	0	0	—	TP	SC
11β OH	S-5	35	40	1	T	SC
11β CH$_3$	S-104	0	0	—	TP	SC
Δ^{14}	S-63	75	50	0.7	T	SC
Δ^{14}	S-63	110-190	125	1	T	O
16α CH$_3$	S-129	<100	<100	1	T	SC
16α F, 16β CH$_3$	S-141	20	0	—	T	SC
16,16 F$_2$	S-100	2.5	—	—	TP	SC
16β CH$_3$	S-130	<100	<100	1	T	SC

Table II.2

Compound	#	Andr	Anab	Q	St	Rte
Parent	S-66	95	75	0.8	T	SC
1α CH$_3$	S-72	15	130	8	TP	SC
1β CH$_3$	S-73	6	50-100	10-15	TP	SC
1α CH$_2$Cl	S-74	5	60	10	TP	SC
1α CH$_2$Br	S-75	1.5	10-20	10	TP	SC
1α CH$_2$I	S-76	1.5	30-40	20	TP	SC
1α, 2α —CH$_2$—	S-78	5	50	10	TP	SC
1,1 (CH$_3$)$_2$	S-77	1.5	2.5	—	TP	SC
1α CH$_3$, 4 Cl	S-79	10	50	5	TP	SC
1α CH$_3$, 7β CH$_3$	S-81	4.5	7.5	—	TP	SC
1α CH$_3$, 7α CH$_3$	S-80	40	400	10	TP	SC
1α CH$_3$, 11β OH	S-82	4.5	2.5	2-4	TP	SC
1α,2α —CH$_2$—, 11β OH	S-83	4.5	50	10	TP	SC
1α CH$_3$, 11 Oxo	S-84	2	20	10	TP	SC
1 CH$_3$, Δ^1	S-85	4.5	7.5	—	TP	SC
1α CH$_3$, Δ^6	S-86	4.5	7.5	—	TP	SC
1 CH$_3$, Δ^1,Δ^6	S-87	4.5	7.5	—	TP	SC
1α CH$_3$, Δ^6, 11 Oxo	S-88	4.5	7.5	—	TP	SC
1α,2α —CH$_2$—, Δ^6	S-89	1.5	10	—	TP	SC
1α CH$_3$, 4 Cl, Δ^6	S-90	4.5	7.5	—	TP	SC
1α,2α —CH$_2$—, 6 Cl, Δ^6	S-91	1.5	6	—	TP	SC
Δ^1, 2 Cl	S-110	5	5	—	T	SC
Δ^1, 4 Cl	S-114	30	80	2.5	TP	SC
2α N(CH$_3$)$_2$	S-41	10	10	—	T	SC
2αN$^+$ (CH$_3$)$_3$ I$^-$	S-42	10	10	—	T	SC
4 Cl	S-146	14	68	4	TP	SC
4 Cl, 11β OH	S-7	25	60	2.5	TP	SC
4 Cl, Δ^6	S-148	0	0	—	TP	SC
4F	S-9	22	25	1	TP	SC
4 OH	S-10	25	65	3	TP	SC
6α Cl	S-29	120	280	2.5	T	SC
6β Cl	S-71	20	20	1	T	SC
6β OAc	S-117	—	<5	—	T	SC
6β NO$_2$	S-39	10	10	—	T	SC
6 Cl, Δ^6	S-64	28	120	4	T	SC
7α SCH$_3$	S-119	140	90	0.6	17 EtNT	SC
7α SCH$_3$	S-119	40	10	0.4	17 EtNT	O
7α SCOCH$_3$	S-120	40	50	1.2	17 EtNT	SC
7α SH	S-121	110	80	0.7	17 EtNT	SC
9α F, 11β OH	S-122	—	35	—	TP	SC

Table II.3

Compound							
	R	#	Andr	Anab	Q	St	Rte
—	Prop	S-3	95	96	1	T	SC
—	Prop	S-3	165	151	~1	T	SC
—	CH_3	S-67	28	23	0.8	T	SC
—	$COCHCl_2$	S-30	350–770	350–770	1	T	SC
—	COCHClF	S-65	297	312	1	T	SC
—	EtO (cyclopentyl)	S-133	190	210	1	MT	O
—	(cyclopentenyl)	S-134	130	110	1	MT	O
—	EtO (cyclohexyl)	S-135	110	220	2	MT	O
—	$Si(CH_3)_3$	S-136	100	100	1	TP	SC
Δ^1	(cyclopentenyl)	S-147	20	60	3	MT	O
$\Delta^1, 9\alpha$ Br, 11β F	Prop	S-124	6	77	12	TP	SC
4 Cl	Prop	S-6	50	100	2	T	SC

Table II.4

Compound	#	Andr	Anab	Q	St	Rte
Parent	S-22	30–40	30–50	1	T	SC
Parent	S-22	50	50	1	MT	O
1α CN	S-37	<10	<10	—	T	SC
1β SCOCH$_3$	S-38	<10	<10	—	T	SC
6α OAc	S-142	0	—	—	—	—
6β OAc	S-143	0	—	—	—	—
6α OH	S-144	0	—	—	—	—
6β OH	S-145	0	—	—	—	—
7α CH$_3$	S-23	45	77	2	T	SC
7α CH$_3$	S-23	65	225	3	T	O
11β OH	S-44	<10	<10	—	T	SC
11 Oxo	S-43	8–50	10–70	1–1.5	T	SC

Table II.5

Compound	#	Andr	Anab	Q	St	Rte
Parent	S-2	94-130	115-150	1.2	T	SC
Parent	S-2	280	280	1	T	O
Parent	S-2	168	159	0.9	TP	SC
1α CH$_3$	S-92	2	20	10	TP	SC
1α SCH$_3$	S-46	82	194	2	MT	O
1α SEt	S-47	35	66	2	MT	O
1α SCOCH$_3$	S-48	33	101	3	MT	O
1α SEt, 7α SEt	S-59	22	122	6	MT	O
1α S Prop, 7α S Prop	S-60	8	40	5	MT	O
1α SCOCH$_3$,7α SCOCH$_3$	S-61	61	456	7	MT	O
1α CH$_3$, 7α CH$_3$	S-94	40	500	12	TP	SC
Δ^1	S-15	50	125	2.5	MT	SC
Δ^1	S-15	40-60	90-210	2-4	MT	O
Δ^1, 2 CH$_3$	S-68	25	50	2	T	SC
Δ^1, 2 CHO	S-137	10	10	1	T	O
Δ^1, 4 Cl	S-105	—	>100	—	MT	O
Δ^1, 7α CH$_3$	S-138	129	332	2.5	MT	O
Δ^1, 7α SH	S-57	59	62	1	MT	O
Δ^1, 7α SCOCH$_3$	S-58	43	75	1.5	MT	O
Δ^1,4 Cl, 11β OH	S-103	<100	<100	1	MT	O
Δ^1, Δ^6, 2 CH$_3$	S-69	120	300	2.5	T	SC
Δ^1,4 Cl, 11β OH	S-103	<100	<100	1	MT	O
Δ^1,9α Cl,11β Cl	S-127	60	300	5	MT	O
Δ^1,9α Br, 11β F	S-128	12	23	2	MT	O
2=CH$_2$	S-32	40	100	2.5	T	SC
2=CH$_2$	S-32	15	50	3	MT	O
2α CH$_3$	S-108	1	23	—	T	SC
3-Dimethylhydrazone	S-116	26	9	0.3	TP	SC
3=CH$_2$	S-26	26	53	2	MT	SC
3=CH$_2$, 9α F, 11β OH	S-28	32	30	1	MT	SC
4 OH	S-16	50	150	3	MT	SC
4 OH	S-16	50	280	6	MT	O
4 Cl	S-17	25	46	2	T	SC
4 Cl	S-17	10	40	4	MT	O
4 SH	S-49	17	18	1	MT	O
4 SCH$_3$	S-50	31	46	1.5	MT	O
4 SEt	S-51	11	24	2	MT	O
4 SCOCH$_3$	S-52	22	47	2	MT	O
4 Cl, 11 Oxo	S-149	6	730	120	Δ^1-MT	O
4 Cl, 11β OH	S-102	>100	>100	1	MT	O

Table II. 5 (*continued*)

Compound	#	Andr	Anab	Q	St	Rte
6α F	S-118	—	230	—	MT	O
6α CH_3	S-96	65	60	0.9	T	SC
6β CH_3	S-97	<100	<100	1	T	SC
Δ^6	S-107	25	60	2.5	MT	O
7α CH_3	S-20	190	146	0.8	T	SC
7α CH_3	S-20	600	1200	2	T	O
7α CH_3	S-20	300	600	2	MT	O
7α SH	S-53	34	144	4	MT	O
7α SCH_3	S-54	7	26	4	MT	O
7α SEt	S-55	20	90	4.5	MT	O
7α $SCOCH_3$	S-56	27	62	2	MT	O
9α Br, 11β F	S-126	4	17	4	MT	O
9α F, 11β OH	S-19	46-188	89-175	1-2	T	SC
9α F, 11β OH	S-19	850	1900	2	MT	O
9α F, 11 Oxo	S-25	850	2200	3	MT	O
9α F, 11β OH, 16α CH_3	S-132	—	14	—	S-19	O
$\Delta^{9\,(11)}$ 16α CH_3	S-131	17	25	1.5	S-19	O
11β OH	S-18	35	40	1.1	MT	SC
11β OH	S-18	90	300	3	MT	O
11 Oxo	S-24	65	140	2	MT	O
$\{17\beta$ OAc, $17\,\alpha CH_3$, Δ^1, $1CH_3\}$	S-93	1.5	2.5	—	TP	SC

Table II.6

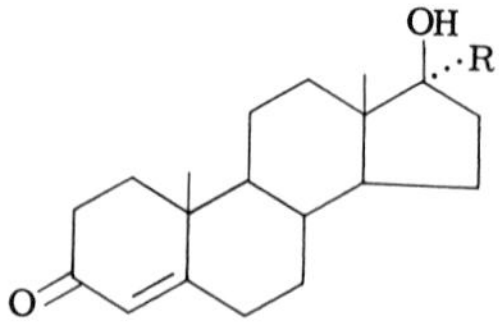

Compound							
	R	#	Andr	Anab	Q	St	Rte
—	Et	S-4	5	10	—	TP	SC
—	Et	S-4	—	30	—	MT	O
$3 = CH_2$	Et	S-27	0.8	4	—	MT	O
$\{11\beta\ OH,\ 17\beta\ OAc\}$	Et	S-99	0	0	—	T	SC
—	C≡CH	S-45	9	4	—	T	SC
$2\alpha\ F$	C≡CH	S-34	15	10	—	T	SC
$2 = CHOH$	C≡CH	S-35	<10	<10	—	T	SC
$\{11\beta\ OH,\ 17\beta\ OAc\}$	C≡CH	S-98	0	0	—	T	SC
$2 = CHOH$	Propynyl	S-36	<10	<10	1	T	SC
—	Propargyl	S-140	0	—	—	T	SC

156

Table III

Number	Structure	Name	V.P.	S.V.	L.A.	Standard	Ref.	Adm.
D-1	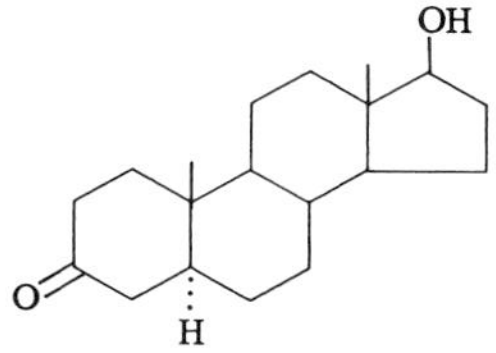	4,5α-Dihydro-testosterone	268	158	152	Testost	30, 61	
			—	30	58	Test prop	36	
			—	85	220	Testost	36	
			120	122	108	Testost	80	
D-2	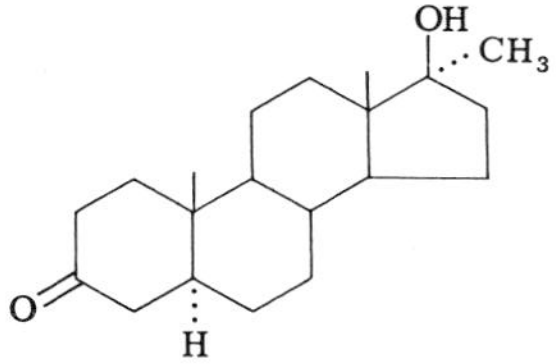	17α-Methyl-4,5α-dihydro-testosterone	254	78	107	Testost	30	
			20	17	24	Test	84	
			64	52	26	17α-MT	84	
			—	20–40	30–40	Test prop	100	
D-3	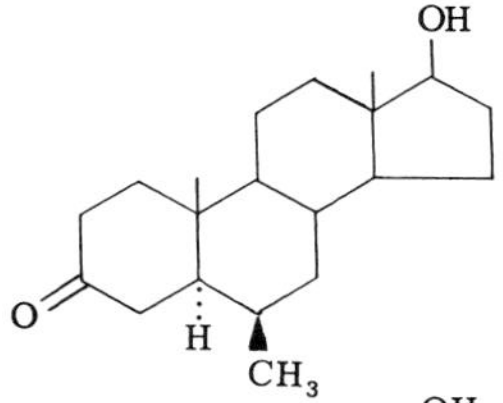	6β-Methyl-4,5α-dihydro-testosterone	390	390	810	Testost	34	
			80	80	73	Testost	107	
D-4	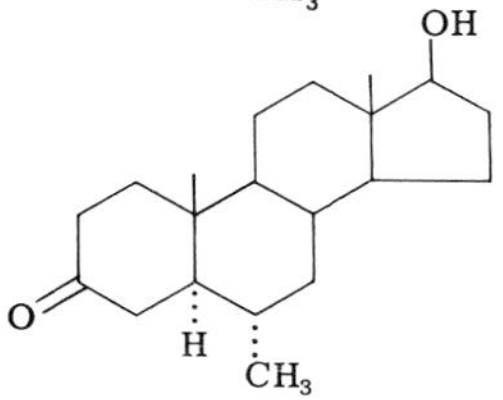	6α-Methyl-4,5α-dihydro-testosterone	23	35	40	Testost	30	
			90	90	460	Testost	34	

157

Table III (*continued*)

Number	Structure	Name	V.P.	S.V.	L.A.	Standard	Ref.	Adm.
D-5	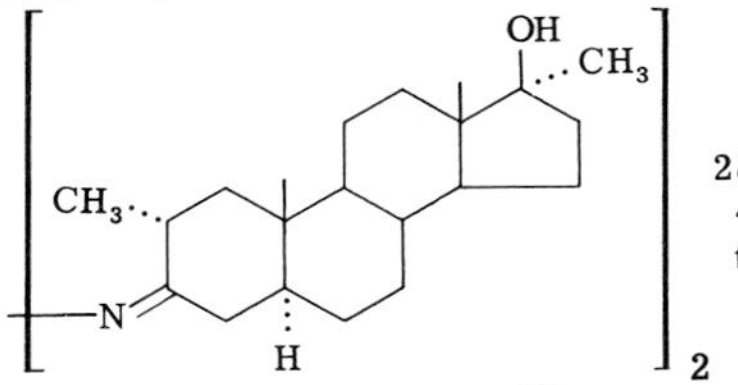 (OH, CH₃, CH₃, N structure — azine ×2)	2α,17α-Dimethyl-4,5α-dihydro-testosterone azine	<6 95	<12 97	27 210	Testost 17α–MT	30 108	
D-6	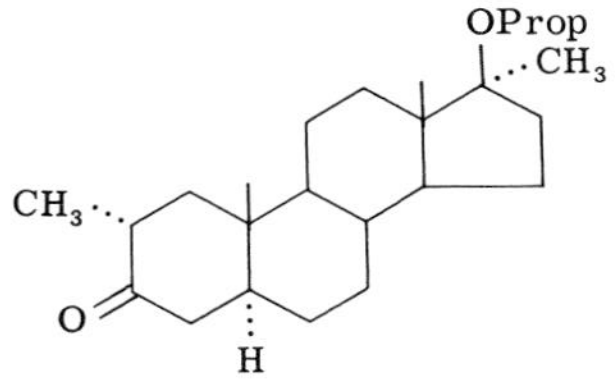 (OProp, CH₃, CH₃, O structure)	2α,17α-Dimethyl-4,5α-dihydro-testosterone propionate	50	50	200	Test prop	39	
D-7	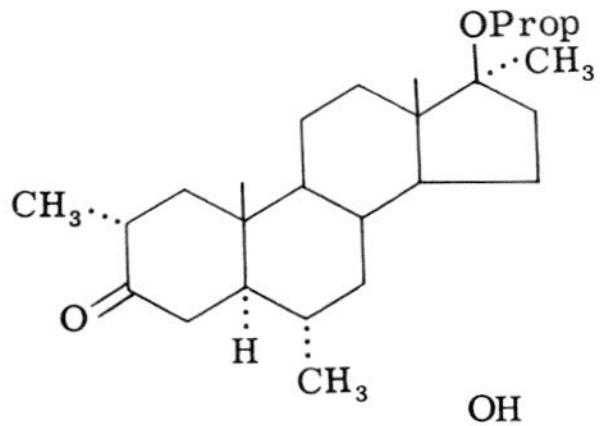 (OProp, CH₃, CH₃, CH₃, O structure)	2α,6α,17α-Trimethyl-4,5α-dihydro-testosterone propionate	50	50	200	Test prop	40	
D-8	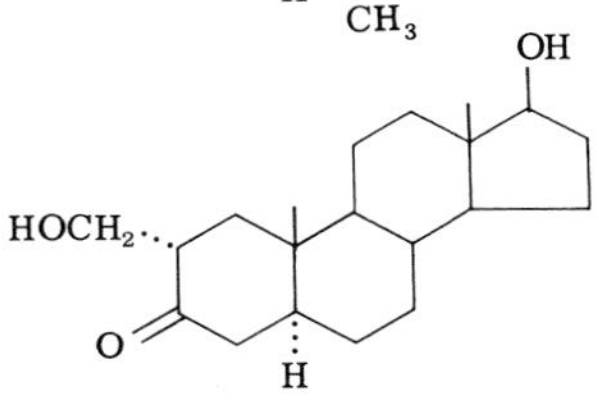 (OH, HOCH₂, O structure)	2α-Hydroxymethyl-4,5α-dihydro-testosterone	35	50	152	Testost	30	

158

	Name				Ref.		Route
D-9	2 - Methylene-17α - methyl-androstan - 17β-ol	37	40	73	Testost	30	
		75	146	205	17α - MT	82	
		100	100	400	17α - MT	57	Oral
D-10	2 - Hydroxymethylene - 17α - methyl-4,5α - dihydro-testosterone	100	64	230	Testost	30	
		30	30	150	17α - MT	9	Oral
		45	45	320	17α - MT	30	Oral
		25	25	400	17α - MT	78	
		36	33	72	Testost	398	
D-11	17α - Methyl - 17β - hydroxy - 5α - androstano[3,2-c]pyrazole	136	169	120	Testost	30	
		33	—	220	17α - MT	124	
		1.65	—	10	Test prop	41	
		47.5	—	196	17α - MT	41	Oral
		69	—	260	—	50	
		3	—	13	Test prop	81	
		30	30	320	17α - MT	26	Oral
D-12	17α - Methyl - 17β - hydroxy - 5α - androstane	—	100	183	17α - MT	42	Oral
		30	30	—	17α - MT	—	

Table III (*continued*)

Number	Structure	Name	V. P.	S. V.	L. A.	Standard	Ref.	Adm.
D-13	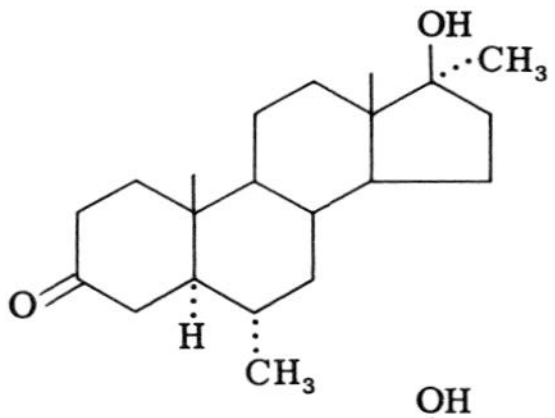	$6\alpha,17\alpha$ - Dimethyl - 17β - hydroxy - 5α - androstan - 3 - one	30	25	50	Testost	56	
D-14	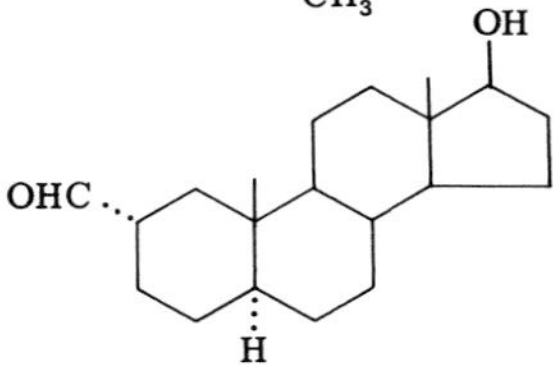	2α - Formyl - 17β - hydroxy - 5α - androstane	<10	<10	40	Testost	56	
D-15	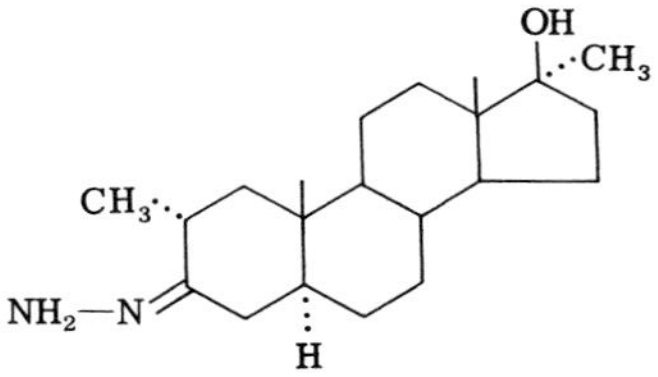	$2\alpha,17\alpha$ - Dimethyl - 5α - androstan - 17β - ol - 3 - hydrazone	79	132	280	17α - MT	108	
D-16	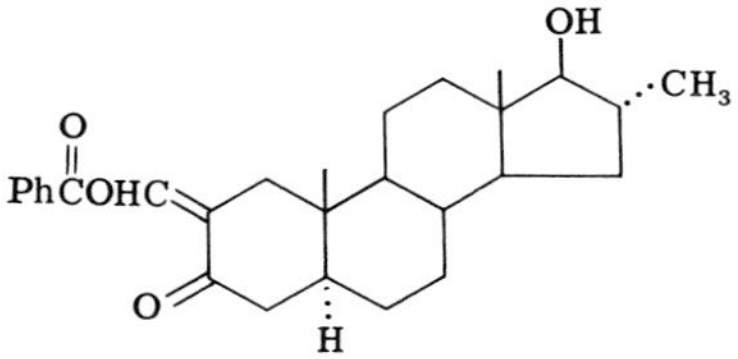	2 - Benzoyloxy-methylene - 17α - methyl - 17β-hydroxy - 5α-androstan - 3 - one	30	20	20	Testost	60	

160

	Compound					
D-17	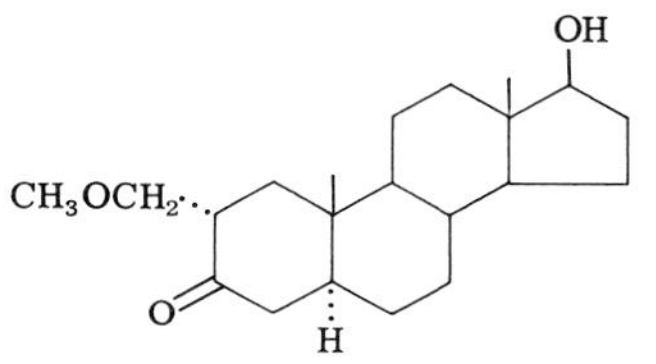 2α-Methoxymethyl-17β-hydroxy-5α-androstan-3-one	10	10	30	Testost	60
D-18	2α-Fluoro-17β-hydroxy-5α-androstan-3-one	20	10	20	Testost	60
		20	20	50	Testost	28
		Weak	Weak	Weak	Testost	125
D-19	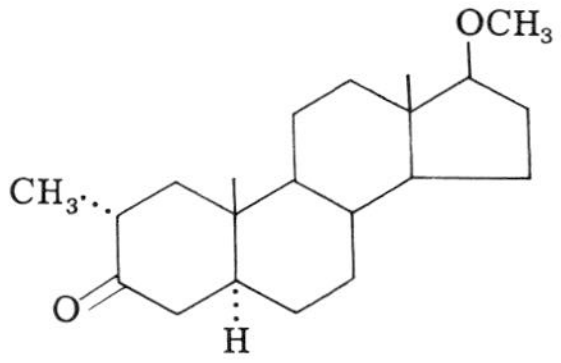 2α-Methyl-17β-methoxy-5α-androstan-3-one	10	10	20	Testost	60
D-20	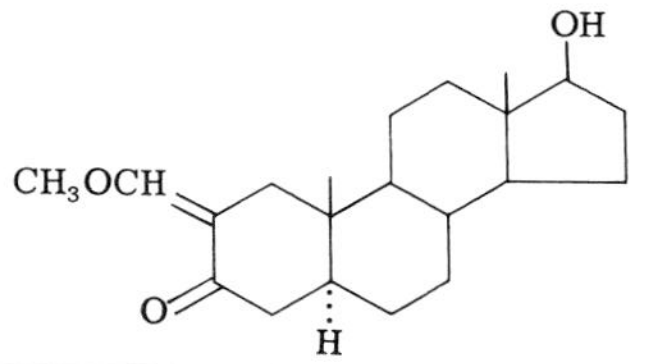 2-Methoxymethylene-17β-hydroxy-5α-androstan-3-one	40	60	40	Testost	60

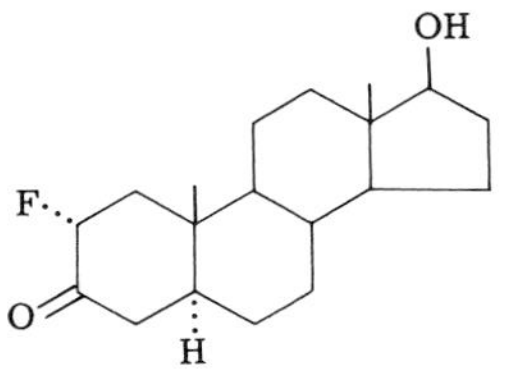

OH

CH$_3$OCH$_2$

D-17

OH

F

D-18

OCH$_3$

CH$_3$

D-19

OH

CH$_3$OCH

D-20

Table III (*continued*)

Number	Structure	Name	V. P.	S. V.	L. A.	Standard	Ref.	Adm.
D-21	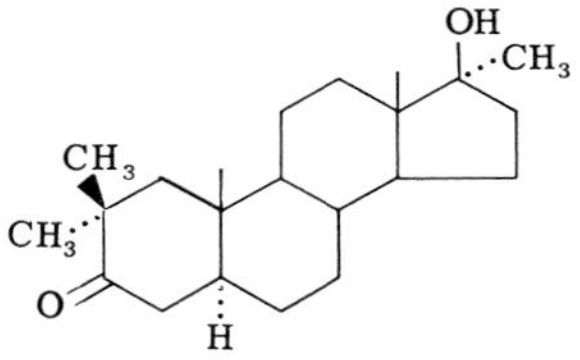	2,2,17α-Trimethyl-17β-hydroxy-5α-androstan-3-one	30	50	20	Testost	60	
D-22	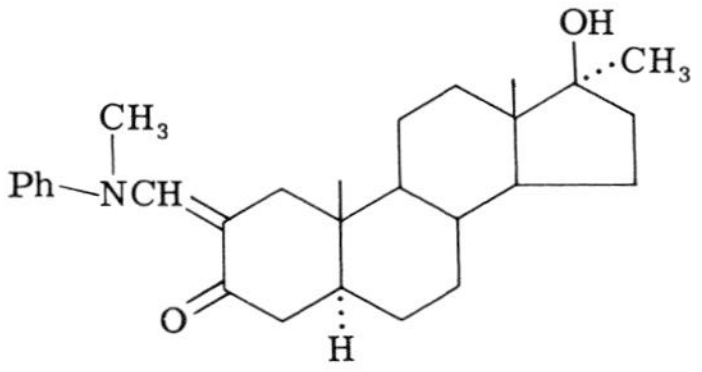	2-N-Methylanilino-methylene-17α-methyl 17β-hydroxy-5α-androstan-3-one	<10	<10	<10	Testost	60	
D-23	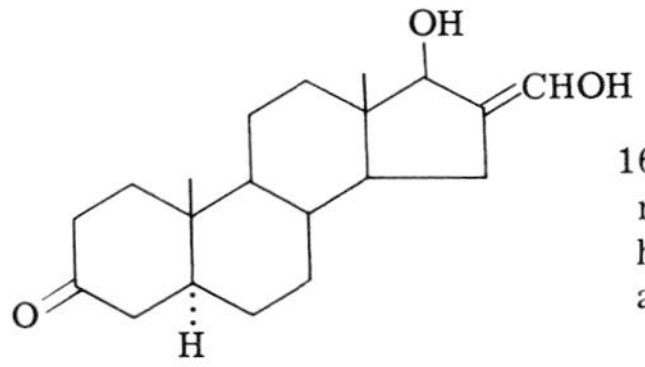	16-Hydroxy-methylene-17β-hydroxy-5α-androstan-3-one	<10	<10	<10	Testost	60	
D-24	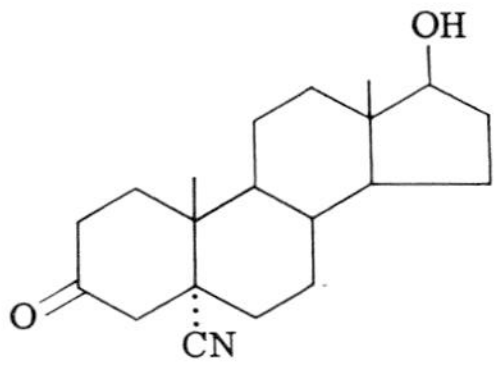	5α-Cyano-17β-hydroxyandrostan-3-one	<10	<10	<10	Testost	60	

162

	Compound					
D-25	5α-Carbamido-17β-hydroxy-androstan-3-one	<10	<10	<10	Testost	60
D-26	5ξ-Methyl-17β-hydroxy-androstan-3-one	30	<5	20	Testost	56
D-27	3α-Hydroxy-5α-androstan-17-one (Androsterone)	53	8	10	Testost	61
		40	24	30	Testost	62
		46*	⟨*By lacrymal gland⟩		Testost	105
		219	129	84	Testost	107
		15	15	—	Testost	165
D-28	2α-Methyl-17β-propionoxy-5α-androstan-3-one	38	32	15	Testost	61
		49	36	114	Testost	82
		50	50	200	Test prop	23
		—	—	166	Testost	107

Table III (continued)

Number	Structure	Name	V. P.	S. V.	L. A.	Standard	Ref.	Adm.
D-29	OAc / AcO / H	3β,17β-Diacetoxy-5α-androstane	13 / 11*	13 / ⟨*By lacrymal gland⟩	8	Testost / Testost	61 / 106	
D-30	O / O / H	5α-Androstan-3,20-dione	33 / 12	13 / —	11 / —	Testost / Testost	61 / 165	
D-31	OH / HO / H	3α,17β-Dihydroxy-5α-androstane	34 / 276 / 283* / 187	24 / 129 / ⟨*By lacrymal gland⟩ / 135	30 / 65 / 110	Testost / Testost / Testost / Testost	61 / 107 / 106 / 398	
D-32	OAc / CH₃ / CH₃ / O / H	1α,7β-Dimethyl-17β-acetoxy-5α-androstan-3-one	—	<4.5	<7.5	Test prop	100	

	Structure	Name				Reference		
D-33		17α-Methyl-17β-hydroxy-5α-androstano[3,2-c]-isoxazole	22	—	155*	17α-MT ⟨*N_2 Retention evaluation⟩	63	Oral
D-34		17β-[(6-Hydroxy methyltetrahydro-pyran-2-yl)oxy]-2α-methyl-5α-androstan-3-one	<100	<100	<100	17α-MT	145	Oral
D-35		2α-Methyl-17β-hydroxy-5α-androstane THP-ether	44-88	67-82	219-395	17α-MT	146	Oral
D-36		2β-3β-Epoxy-17β-acetoxy-5α-androstane	<10	<10	10	Test prop	71	

165

Table III (*continued*)

Number	Structure	Name	V. P.	S. V.	L. A.	Standard	Ref.	Adm.
D-37	OAc ... O ... H	2α,3α-Epoxy-17β-acetoxy-5α-androstane	<10	<10	<20	Test prop	71	
D-38	OH ...CH₃ ... O ... H	2α,3α-Epoxy-17α-methyl-17β-hydroxyandrostane	<20 0 1.5	<20 0 1.5	20 0 5	Test prop 17α-MT Test prop	71 87 87	Oral
D-39	OAc ... O ... H	3α,4α-Epoxy-17β-acetoxy-5α-androstane	<20	<20	20	Test prop	71	
D-40	OH ... H	2α,3α-Methano-17β-hydroxy-5α-androstane	30	30	100	Test prop	71	

						Oral
D-41		2α,3α-Methano-17α-methyl-17β-hydroxy-5α-androstane	<10	<10	10	Test prop	71
D-42		Spiro-2β-oxiranyl-17β-hydroxy-5α-androstane	0	0	0	Test prop	71
D-43		Spiro-3β-oxiranyl-17β-hydroxy-5α-androstane	<10 / 0	<10 / 0	<10 / 0	Test prop / Test prop	71 / 54
D-44		2α-Acetylthio-17α-methyl-17β-hydroxy-5α-androstan-3-one	—	7	6	17α-MT	76

Table III (continued)

Number	Structure	Name	V.P.	S.V.	L.A.	Standard	Ref.	Adm.
D-45		2α,17α-Dimethyl-17β-hydroxy-5α-androstan-3-one	20 —	20 —	400 800	17α-MT 17α-MT	78 107	Oral
D-46		4β-Methyl-17β-hydroxy-5α-androstan-3-one	10	10	200	Testost	78	
D-47		4α-Methyl-17β-hydroxy-5α-androstan-3-one	10 —	10 —	100 50	Testost Testost	78 107	
D-48		4,5α-Dihydro-Δ^{14} testosterone	83 142	28.5 120	55 143	Testost Testost	80 80	Oral

D-49	2α - Methyl - 17β - hydroxy - 5α - androstan - 3 - one	24	26	62	Testost	82
		—	40	130	Testost	107
D-50	2α - Methyl - 17β - hydroxy - 5α - androst - 9(11) - en - 3 - one	62	68	150	Testost	82
D-51	2α,3α - Difluoro - methylene - 5α - androstan - 17β - ol - acetate	27	44	122	Testost	82
D-52	2α,3α - Difluoro - methylene - 17α - methyl - 5α - androstan - 17β - ol	5	11	28	Testost	82
		50	—	250	17α - MT	82 Oral

Table III (continued)

Number	Structure	Name	V.P.	S.V.	L.A.	Standard	Ref.	Adm.
D-53		$2\beta,3\beta$-Difluoro-methylene-5α-androstan-17β-ol acetate	34	42	84	Testost	82	
D-54		3-Methylene-17α-methyl-17β-hydroxy-5α-androstane	84 —	32 11	69 59	17α-MT 17α-MT	82 48	Oral Oral
D-55		$2\alpha,17\beta$-Dimethyl-17β-hydroxy-5α-androst-9(11)-en-3-one	62	65	153	Testost	82	
D-56		17α-Methyl-17β-hydroxy-5α-androstan-1-one	7.7 180	6.7 190	35 830	Test prop 17α-MT	84 84	Oral

	Structure	Compound				Standard	Ref.	Route
D-57		17α-Methyl-17β-hydroxy-5α-androstan-2-one	<0.5 / <3	<0.4 / <5	2.0 / 16	Test prop / 17α-MT	84 / 84	Oral
D-58		17α-Methyl-17β-hydroxy-5α-androstan-4-one	0.6 / <20	0.6 / <24	1.8 / <35	Test prop / 17α-MT	84 / 84	Oral
D-59		17α-Methyl-1α,17β-Dihydroxy-5α-androstane	6.2 / 170	4.2 / 150	23.0 / 660	Test prop / 17α-MT	84 / 84	Oral
D-60		17α-Methyl-2β,17β-Dihydroxy-5α-androstane	0.5 / <6	0.4 / <6	1.5 / <13	Test prop / 17α-MT	84 / 84	Oral

171

Table III (continued)

Number	Structure	Name	V.P.	S.V.	L.A.	Standard	Ref.	Adm.
D-61		17α-Methyl-3β,17β-dihydroxy-5α-androstane	4.0 72	4.3 65	1.6 22	Test prop 17α-MT	84 84	Oral
D-62		17α-Methyl-4β-17β-dihydroxy-5α-androstane	<0.5 <20	<0.4 <24	<0.1 <35	Test prop 17α-MT	84 84	Oral
D-63		17α-Methyl-5α-androstane	—	50	280	17α-MT	86	Oral
D-64		2α,3α-Epoxy-17β-hydroxy-5α-androstane	0.5 0	0.5 0	0.8 0	Test prop 17α-MT	87 87	Oral

No.	Structure	Name				Standard	Ref.	Route
D-65		2β,3β - Epithio - 17β - hydroxy - 5α - androstane	2.2 0	2.2 0	13 0	Test prop 17α – MT	87 87	Oral
D-66		2β,3β - Epithio - 17α - methyl - 17β - hydroxy - 5α - androstane	0.6 18	0.6 18	2.8 78	Test prop 17α – MT	87 87	Oral
D-67		2β,3β - Epoxy - 17β - hydroxy - 5α - androstane	1.3 0	1.3 0	6.7 0	Test prop 17α – MT	87 87	Oral
D-68		2β,3β - Epoxy - 17α - methyl - 17β - hydroxy - 5α - androstane	5.6 0	5.6 0	36 0	Test prop 17α – MT	87 87	Oral

173

Table III (continued)

Number	Structure	Name	V. P.	S. V.	L. A.	Standard	Ref.	Adm.
D-69		2α,3α-Epithio-17β-hydroxy-5α-androstane	42 25	42 25	308 100	Test prop 17α-MT	87 87	Oral
D-70		2α,3α-Epithio-17α-methyl-17β-hydroxy-5α-androstane	27 91	27 91	154 1100	Test prop 17α-MT	87 87	Oral
D-71		17α-Methyl-5α-androstano[2,3-c][1'2'5']oxadiazol-17β-ol (or -[2,3-c]furazan)	<100 19-26 73-94	<100 19-26 73-94	<100 100-150 270-330	Testost Test prop 17α-MT	90 121 121	
D-72		17α-Methyl-17β-hydroxy-2β(H)-2,3-cyclohex-2'-eno-5α-androstan-4'-one	0	0	0	Testost	150	

	Structure	Name				Standard	
D-73		17α - Methyl - 17β - hydroxy - 2,3 - cyclohexano - 5α - androstan - 4' - one	0	0	0	Testost	151
D-74		16α - Fluoro - 16β - methyl - 17β - hydroxy - 5α - androstan - 3 - one	10	30	0	Testost	96
D-75		2 - Methylene - 3β, 17β - diacetoxy - 5α - androstane	—	1	<5	Test prop	97
D-76		17α - Methyl- 2 - methylene - 3β, 17β - dihydroxy - 5α - androstane	—	<1	<5	Test prop	97

Table III (*continued*)

Number	Structure	Name	V.P.	S.V.	L.A.	Standard	Ref.	Adm.
D-77		2 - Methylene - 3β - acetoxy - 17α - methyl - 17β - hydroxy - 5α - androstane	— —	1 <5	5 <5	Test prop 17α - MT	97 97	Oral
D-78		2 - Aminomethylene - 17α - methyl - 17β - hydroxy - 5α - androstan - 3 - one	20	20	160	17α - MT	94	Oral
D-79		2[2' - (N,N - Dimethyl - amino)ethylamino - methylene] - 17α - methyl - 5α - androstan - 17β - ol - 3 - one	30	30	150	17α - MT	94	Oral
D-80		2[2' - (N,N - Diethyl - amino)ethylamino - methylene] - 17α - methyl - 5α - androstan - 17β - ol - 3 - one	40 40	40 40	65 100	Testost 17α - MT	94 94	Oral

D-81		2 - N - Piperidyl - methylene - 17α - methyl - 17β - hydroxy - 5α - androstan - 3 - one	30 <50	30 <50	60 <50	Testost 17α - MT	94 94
D-82		2 - N,N - Dimethyl- aminomethylene - 17α - methyl - 5α - androstan - 17β - ol - 3 - one	<10 20	<10 20	<10 100	Testost 17α - MT	94 94
D-83		2 - N,N - Diethyl- aminomethylene - 17α - methyl - 5α - androstan - 17β - ol - 3 - one	20	20	120	17α - MT	94
D-84		1α - Methyl - 17β - hydroxy - 5α - androstan - 3 - one	— 77	30-40 73	100-150 136	Test prop Testost	100 398 (See also 406)

177

Table III (continued)

Number	Structure	Name	V. P.	S. V.	L. A.	Standard	Ref.	Adm.
D-85		1β - Methyl - 17β - hydroxy - 5α - androstan - 3 - one	—		<1.5	5-10	Test prop	100	
D-86		1 - Methylene - 17β - hydroxy - 5α - androstan - 3 - one	—	5-10	20		Test prop	100	
D-87		1ξ - Formyl - 17β - hydroxy - 5α - androstan - 3 - one	—		<1.5	<2.5	Test prop	100	
D-88		1α - Cyano - 17β - hydroxy - 5α - androstan - 3 - one	—		<4.5	<7.5	Test prop	100	

178

	Name					
D-89	1α,17β-Dihydroxy-5α-androstan-3-one	—	50	200-400	Test prop	100
D-90	1ξ-Isothiocyano-17β-hydroxy-5α-androstan-3-one	—	10	50-100	Test prop	100
D-91	1α-Hydroxymethyl-17β-hydroxy-5α-androstan-3-one	—	<4.5	<7.5	Test prop	100
D-92	1ξ-Nitromethyl-17β-hydroxy-5α-androstan-3-one	—	<4.5	<7.5	Test prop	100

Table III (continued)

Number	Structure	Name	V.P.	S.V.	L.A.	Standard	Ref.	Adm.
D-93		1,1 - Dimethyl - 17β - hydroxy - 5α - androstan - 3 - one	—	1.5	2.5	Test prop	100	
D-94		1,1 - Ethylene - 17β - hydroxy - 5α - androstan - 3 - one	—	1.5	6	Test prop	100	
D-95		Spiro - 1α - oxiranyl - 17β - hydroxy - 5α - androstan - 3 - one	—	<4.5	<7.5	Test prop	100	
D-96		1α,7α - Dimethyl - 17β - hydroxy - 5α - androstan - 3 - one	—	100	200-400	Test prop	100	

D-97	1,2α - Methylene - 4 hydroxymethylene - 17β - hydroxy - 5α - androstan - 3 - one	—	3	20-30	Test prop	100
D-98	Δ² - Pyrazolino - 4',3', 1,2 - androstan - 17β - ol - 3 - one	—	<1.5	5-10	Test prop	100
D-99	17β - Acetoxy - 5α - androstan - 3 - one	—	30	100-200	Test prop	100
D-100	1α - Methyl - 17β - acetoxy - 5α - androstan - 3 - one	—	40	200-300	Test prop	100

181

Table III (*continued*)

Number	Structure	Name	V. P.	S. V.	L. A.	Standard	Ref.	Adm.
D-101		1β – Methyl – 17β – acetoxy – 5α – androstan – 3 – one	—	1.5	10	Test prop	100	
D-102		1α – Isopropyl – 17β – acetoxy – 5α – androstan – 3 – one	—	<4.5	<7.5	Test prop	100	
D-103		1 – Methylene – 17β – acetoxy – 5α – androstan – 3 – one	— 10 4	10 10 9	30 20 15	Test prop Testost Testost	100 56 82	
D-104		1α– Chloromethyl – 17β – acetoxy – 5α– androstan – 3 – one	—	2-3	10-20	Test prop	100	

	Structure	Name					
D-105	CH₂Br, OAc, H, O (1α-CH$_2$Br, 17β-OAc)	1α – Bromomethyl – 17β – acetoxy – 5α – androstan – 3 – one	—	4	30	Test prop	100
D-106	CH₂I, OAc, H, O	1α – Iodomethyl – 17β – acetoxy – 5α – androstan – 3 – one	—	3	20	Test prop	100
D-107	OH, OAc, H, O	1α – Hydroxy – 17β – acetoxy – 5α – androstan – 3 – one	—	70	200– 300	Test prop	100
D-108	OAc, OAc, H, O	$1\alpha,17\beta$ – Diacetoxy – 5α – androstan – 3 – one	—	70	200– 400	Test prop	100

Table III (*continued*)

Number	Structure	Name	V.P.	S.V.	L.A.	Standard	Ref.	Adm.
D-109	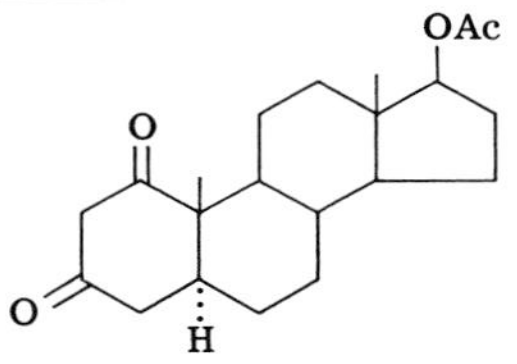	17β – acetoxy – 5α – androstan – 1,3 – dione	—	13	<100	Test prop	100	
D-110	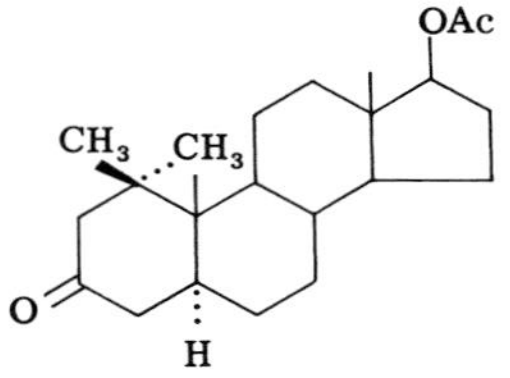	1,1 – Dimethyl – 17β – acetoxy – 5α – androstan – 3 – one	—	<4.5	<7.5	Test prop	100	
D-111	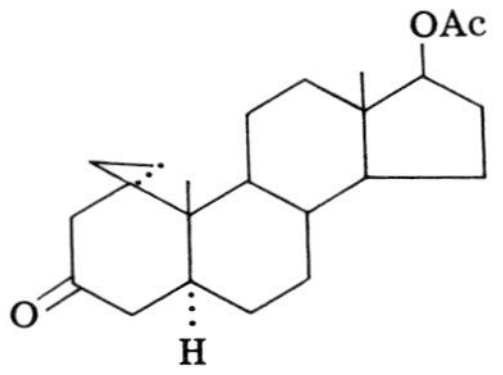	1,1 – Ethylene – 17β – acetoxy – 5α – androstan – 3 – one	—	<4.5	<7.5	Test prop	100	
D-112	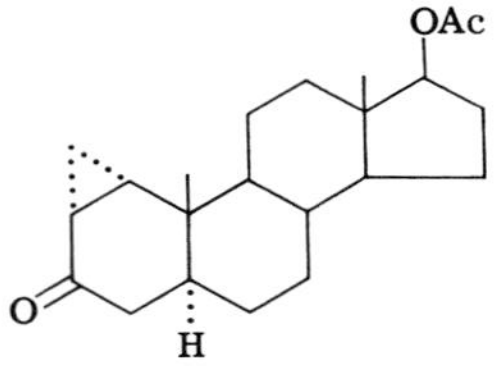	1,2α – Methylene – 17β – acetoxy – 5α – androstan – 3 – one	—	15	50–100	Test prop	100	

184

D-113	1,2α - Methylene - 17β-acetoxy - 4 - hydroxy - methylene - 5α - androstan - 3 - one	—	<4.5	<7.5	Test prop	100
D-114	1,2α - Methylene - 17β-acetoxy - 11β - hydroxy - 5α - androstan - 3 - one	—	<4.5	<7.5	Test prop	100
D-115	Spiro - 1α - oxiranyl - 17β - acetoxy - 5α - androstane - 3 - one	—	<4.5	<7.5	Test prop	100
D-116	1,2ξ - Dimethyl - 17β - acetoxy - 5α-androstan - 3 - one	—	<15	<25	Test prop	100

Table III (*continued*)

Number	Structure	Name	V.P.	S.V.	L.A.	Standard	Ref.	Adm.
D-117	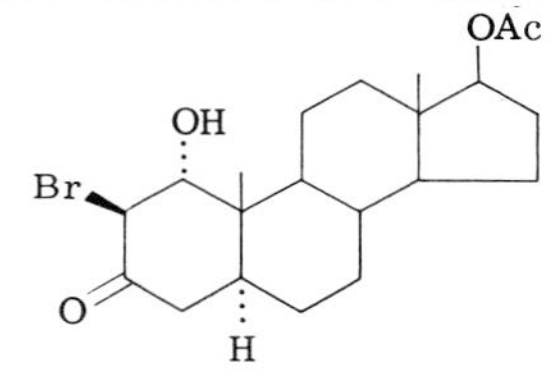	2β - Bromo-1α - hydroxy - 17β - acetoxy - 5α - androstan - 3 - one	—	2	10–20	Test prop	100	
D-118	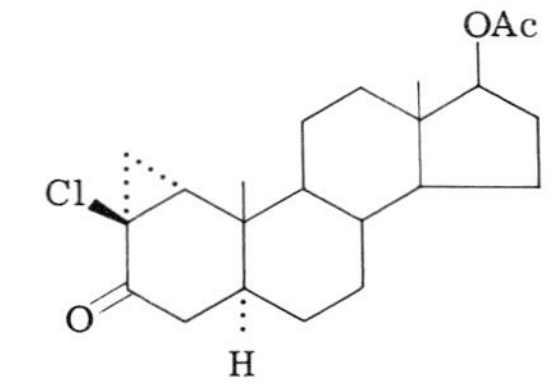	2β- Chloro - 1α,2α - methylene - 17β - acetoxy - 5α - androstan-3-one	—	<4.5	<7.5	Test prop	100	
D-119	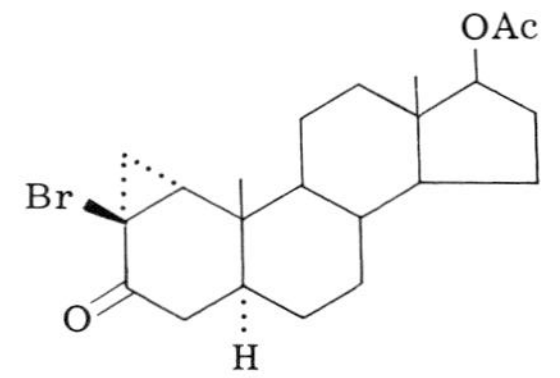	2β - Bromo-1,2α - methylene - 17β - acetoyx-5A - androstan-3-one	—	<4.5	<7.5	Test prop	100	
D-120	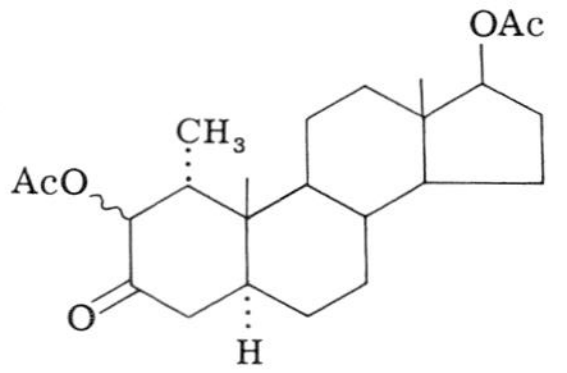	1α - Methyl - 2ξ, 17β - diacetoxy - 5α - androstan - 3 - one	—	<4.5	<7.5	Test prop	100	

186

	Structure	Name				
D-121		1 - Methylene - 2,2 - dimethyl1 - 17β - acetoxy - 5α - androstan - 3 - one	—	<4.5	<7.5	Test prop 100
D-122		1α,7α - Dimethyl - 17β - acetoxy - 5α - androstan - 3 - one	—	50-100	50-100	Test prop 100
D-123		1α,17α - Dimethyl - 17β - hydroxy - 5α - androstan - 3 - one	—	20-30	150	Test prop 100
D-124		1β,17α - Dimethyl - 17β - hydroxy - 5α - androstan - 3 - one	—	1.5	10	Test prop 100

Table III (continued)

Number	Structure	Name	V. P.	S. V.	L. A.	Standard	Ref.	Adm.
D-125	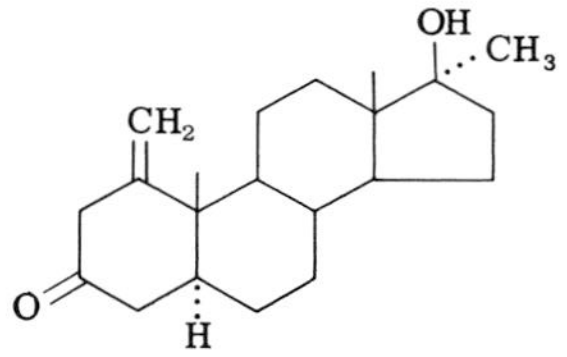	1 - Methylene - 17α - methyl - 17β - hydroxy - 5α - androstan - 3 - one	—	15	30	Test prop	100	
D-126	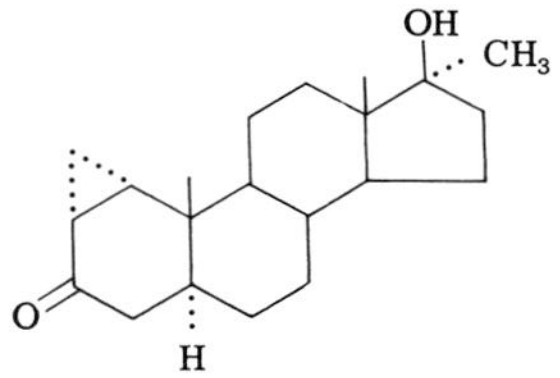	1,2α- Methylene - 17α- methyl - 17β- hydroxy - 5α- androstan - 3 - one	—	10	30	Test prop	100	
D-127	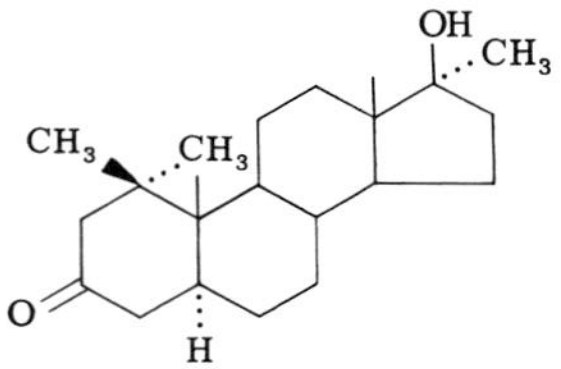	1,1,17α- Trimethyl - 17β - hydroxy - 5α- androstan - 3 - one	—	1.5	10	Test prop	100	

188

				Standard	
D-128	$1\alpha,7\alpha,17\alpha$-Trimethyl-5α-androstan-17β-ol-3-one	—	60	200-400	Test prop	100
D-129	1α-Methyl-17β-hydroxy-5α-androstane	—	<1.5	5	Test prop	100
D-130	5α-Androstane 17β-ol-3-one (1'-methoxy)cyclopentyl ether	122	270	308	17α-MT	143	Oral
D-131	$1,2\alpha$-Methylene-5α-androstan-17β-ol	—	<4.5	20-30	Test prop	100

189

Table III (*continued*)

Number	Structure	Name	V. P.	S. V.	L. A.	Standard	Ref.	Adm.
D-132		5α-Androstane-17β-ol-3-one 17-(1'-ethoxy)cyclopentyl ether	130	270	300	17α-MT	144	Oral
D-133		2α-Methyl-3β,17β-dihydroxy-5α-androstane	—	—	16	Testost	107	
D-134		17β-Acetoxy-5α-androstane	—	<1.5	3.0	Test prop	100	
D-135		1α-Methyl-17β-acetoxy-5α-androstane	—	<1.5	<2.5	Test prop	100	

D-136	1β - Methyl - 17β - acetoxy - 5α - androstane	—	<1.5	<2.5	Test prop	100
D-137	1,2α - Methylene - 17β - acetoxy - 5α - androstane	—	2.1	22	Test prop	100
D-138	17β - Acetoxy - 5α - androstan - 1 - one	—	9	40	Test prop	100
D-139	1α,17α - Dimethyl - 17β - hydroxy - 5α - androstane	—	<4.5	7.5	Test prop	100

191

Table III (*continued*)

Number	Structure	Name	V.P.	S.V.	L.A.	Standard	Ref.	Adm.
D-140	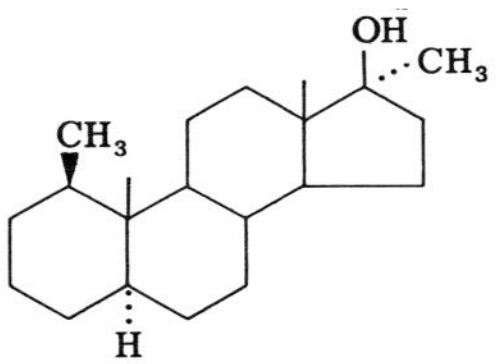	1β,17α-Dimethyl-17β-hydroxy-5α-androstane	—	<1.5	2.5	Test prop	100	
D-141	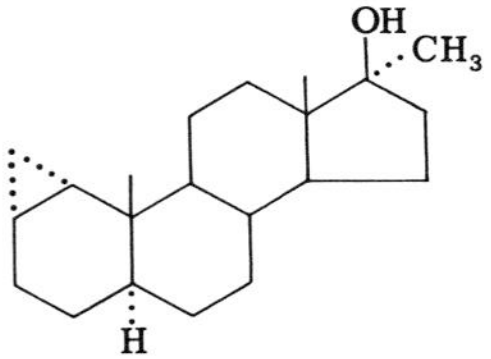	1,2α-Methylene-17α-methyl-17β-hydroxy-5α-androstane	—	4	30	Test prop	100	
D-142	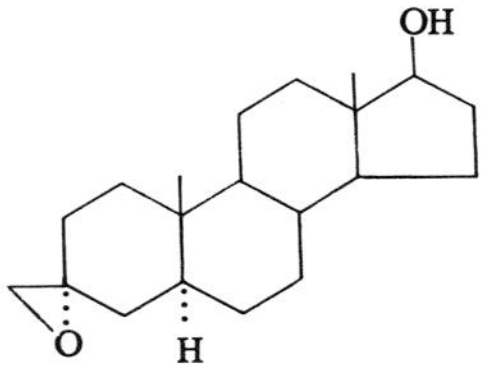	Spiro-3α-oxiranyl-5α-androstan-17β-ol	0	0	0	Test prop	54	
D-143	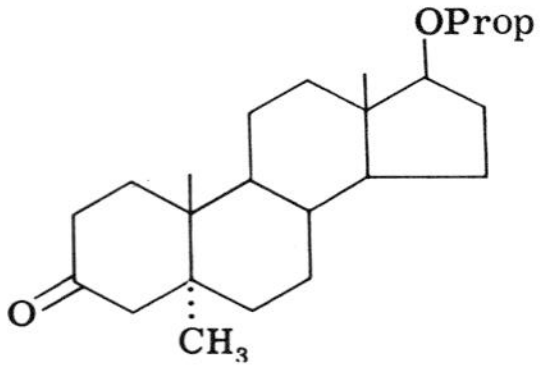	5α-Methyl-17β-propionoxy-androstan-3-one	0	0	0	Testost	132	

192

Structure	Name	Increase in weight of chick comb 1.5% of Testost (intramuscular injection or inunction onto comb.)		
D-144	5α-Androstane		22	
D-145	5α-Androstan-17β-ol	4.8* / 22.5	⟨By lacrymal gland⟩ *22.5 / 18	Testost 106 / Testost 107
D-146	3β,17β-Dihydroxy-5α-androstane	4.9* / 3	⟨By lacrymal gland⟩ 3 / —	Testost 106 / Testost 165
D-147	2-Hydroxymethylene-17β-hydroxy-5α-androstan-3-one	—	30	Testost 107

Table III (continued)

Number	Structure	Name	V. P.	S. V.	L. A.	Standard	Ref.	Adm.
D-148		4,5α - Epoxido - 11β, 17β - dihydroxy - 17α - methylandrostan - 3 - one	⟨ Less effective ⟩			Δ¹ - 17α - MT	109	
D-149		17α - Methyl - 17 - β - hydroxyandrostano - [3,2-b]pyridine	Weak	Weak	Weak	Test prop	122	
D-150		3,3 - Azo - 17α - methyl - 5α - androstan - 17β - ol	20	—	300	17α - MT	123	Oral
D-151		2α - Formyl - 17β - methyl - 17β - hydroxy - 5α - androstan - 3 - one	—	—	100	17α - MT	107	

D-152	2 - Methylene - 17α - methyl - 17β - hydroxy - 5α - androstan - 3 - one	15	15	50	17α – MT	98
D-153	2β - Fluoro - 17β - hydroxy - 5α - androstan - 3 - one	0	0	0	Testost	125
D-154	4,5α - Dihydro- testosterone - 3 - iso- nicotinyl hydrazone	35– 130	68– 133	52– 98	Testost	30
D-155	17β - Hydroxy - 5α - androstane - 3 - fulvene	<5	<5	<5	Testost	107

Table III (continued)

Number	Structure	Name	V.P.	S.V.	L.A.	Standard	Ref.	Adm.
D-156		3β-Cyclopentyl-17β-hydroxy-5α-androstane	<5	<5	<5	Testost	107	
D-157		2-Methylene-17β-hydroxy-5α-androstane	10	10	20	Testost	57	
D-158		2α-Methyl-17β-dichloroacetoxy-5α-androstan-3-one	71	84	206	Testost	82	
D-159		11α,17α-Dimethyl-11β,17β-dihydroxy-5α-androstan-3-one	0	0	0	Testost	161	

		Name				
D-160		11α,17α-Dimethyl-3β,11β,17β-trihydroxy-5α-androstane	0	0	0	Testost	161
D-161		Epiandrosterone	2	2	—	Testost	165
D-162		3β,17β-Dihydroxy-3α-methyl-5α-androstane	0	0	—	Testost	3
D-163		3α,17β-Dihydroxy-3β-methyl-5α-androstane	0	0	—	Testost	3

197

Table III *(continued)*

Number	Structure	Name	V. P.	S. V.	L. A.	Standard	Ref.	Adm.
D-164	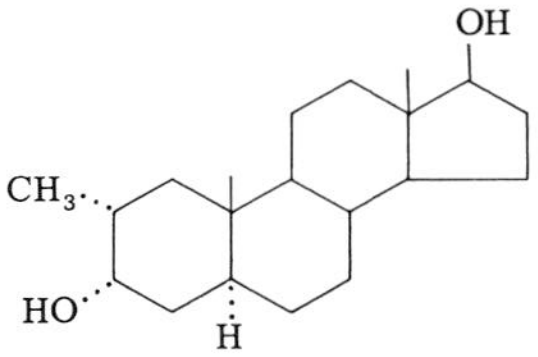	2α – Methyl – 3α,17β – dihydroxy – 5α – androstane	—	—	25	Testost	107	
D-165	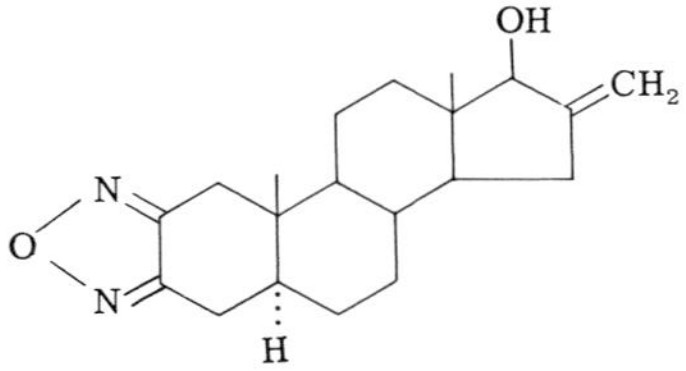	16 – Methylene – 17β – hydroxy – 5α – androstano [2,3-c] furazan	Weak	Weak	Weak	D-169	395	
D-166	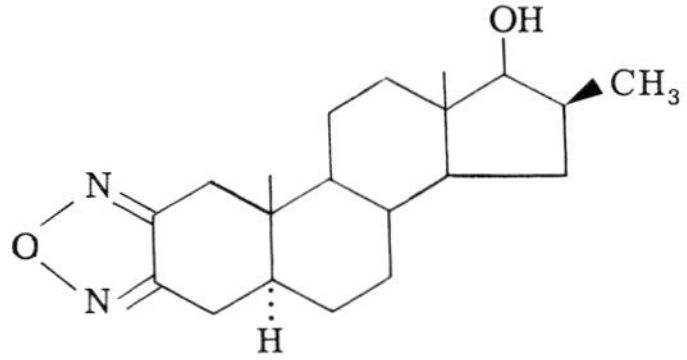	16β – Methyl – 17β – hydroxy – 5α – androstano [2,3-c] furazan	Weak	Weak	Weak	D-169	395	
D-167	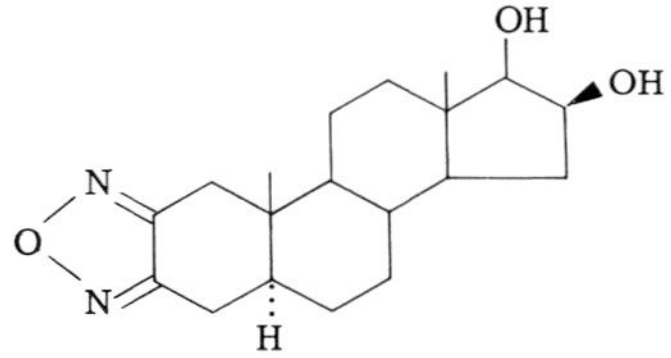	16β,17β – Dihydroxy – 5α – androstano [2,3-c] furazan	Weak	Weak	Weak	D-169	395	

198

D-168	16α,17β-Dihydroxy-5α-androstano-[2,3-c]furazan	Weak	Weak	Weak	D-169	395
D-169	17β-Hydroxy-5α-androstano[2,3-c]furazan	50	30	~100	Test prop	394
D-170	1α-Ethyl-17β-hydroxy-5α-androstan-3-one	0	0	0	Testost	398
D-171	1α-Methyl-3α,17β-dihydroxy-5α-androstane	112	97	164	Testost	398

Table III (*continued*)

Number	Structure	Name	V. P.	S. V.	L. A.	Standard	Ref.	Adm.
D-172	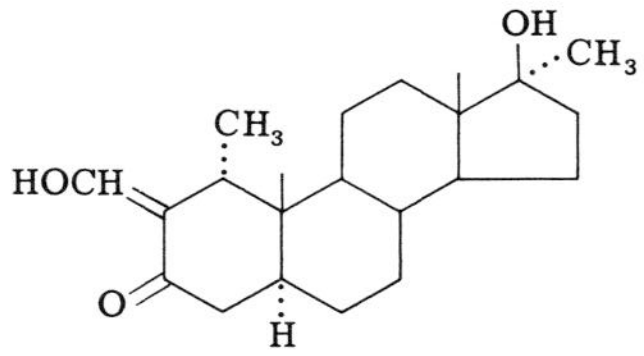	1α,17α - Dimethyl - 17β - hydroxy - 2 - hydroxymethylene - 5α - androstan - 3 - one	24	24	42	Testost	398	
D-173	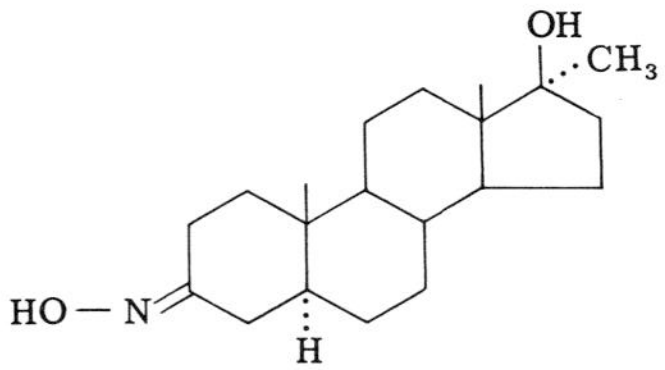	17β - Hydroxy - 17α - methyl - 5α - androstan - 3 - one oxime	158	117	380	Testost	398	
D-174	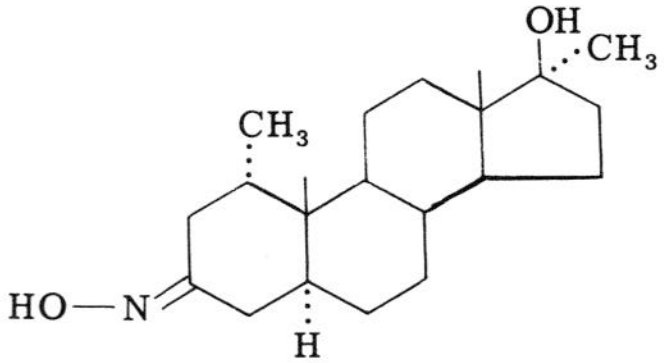	17β - Hydroxy - 1α, 17α - dimethyl - 5α - androstan - 3 - one oxime	71	64	170	Testost	398	
D-175	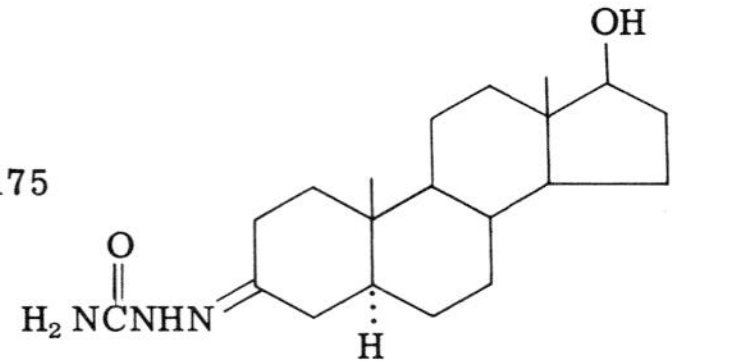	17β - Hydroxy - 5α - androstan - 3 - one semicarbazone	64	53	76	Testost	398	

200

D-176	1α-Methyl-17β-hydroxy-5α-androstan-3-one semicarbazone	23	24	60	Testost	398
D-177	17β-Hydroxy-1α,17β-dimethyl-5α-androstano[3,2-c]pyrazole	28	30	59	Testost	398

Table III.1

Compound	#	Andr	Anab	Q	St	Rte
Parent	D-1	30-260	60-220	2	TP,T	SC
1α CH₃	D-84	30-40	100-150	3-5	TP	SC
1α Et	D-170	0	0	—	T	SC
1α CN	D-88	4.5	7.5	—	TP	SC
1α OH	D-89	50	200-400	4-8	TP	SC
1α CH₂ OH	D-91	4.5	7.5	—	TP	SC
1α CH₂ NO₂	D-92	4.5	7.5	—	TP	SC
1β CH₃	D-85	1.5	5-10	—	TP	SC
1ξ CHO	D-87	1	2	—	TP	SC
1ξ SCN	D-90	10	50-100	5-10	TP	SC
1 =CH₂	D-86	5-10	20	2-4	TP	SC
1,1 (CH₃)₂	D-93	1.5	2.5	—	TP	SC
1,1 ▽	D-94	1.5	6	—	TP	SC
1,1 O▽	D-95	4.5	7.5	—	TP	SC
1α,2α —CH₂ —						
4 =CHOH	D-97	3	30	10	TP	SC
1,2 - Pyrazolino	D-98	1.5	5-10	—	TP	SC
1α CH₃,7α CH₃	D-96	100	200-400	4	TP	SC
2α CH₃	D-49	25	62	3	T	SC
2α CH₂ OH	D-8	40-50	150	3	T	SC
2α CH₂ OCH₃	D-17	10	30	3	T	SC
2α CH₃, Δ⁹⁽¹¹⁾	D-50	65	150	2.8	T	SC
2α F	D-18	20	20-50	1-2	TP	SC
2β F	D-153	0	0	—	TP	SC
2 =CHOH	D-147	—	30	—	T	SC
2 =CHOCH₃	D-20	50	40	0.8	TP	SC
4α CH₃	D-47	10	100	10	TP	SC
4β CH₃	D-46	10	200	20	TP	SC
5α CH₃	D-26	5-30	20	< 1	TP	SC
5α CN	D-24	10	10	1	TP	SC
5α CONH₂	D-25	10	10	1	TP	SC
6α CH₃	D-4	30	40	1.2	TP	SC
6β CH₃	D-3	80	100	1.2	TP	SC
Δ¹⁴⁽¹⁵⁾	D-48	50	50	1	TP	SC
16 =CHOH	D-23	10	10	1	TP	SC
16α F, 16β CH₃	D-74	20	0	—	TP	SC

Table III.2

Compound	#	Andr	Anab	Q	St	Rte
Parent	D-2	20-40	30-40	1	TP	SC
Parent	D-2	55	26	0.5	MT	O
1α CH$_3$	D-123	25	150	6	TP	SC
1β CH$_3$	D-124	1.5	10	—	TP	SC
1,1 (CH$_3$)$_2$	D-127	1.5	10	—	TP	SC
1=CH$_2$	D-125	15	30	2	TP	SC
$1\alpha,2\alpha-$CH$_2-$	D-126	10	30	3	TP	SC
1α CH$_3$,2=CHOH	D-172	24	42	2	T	SC
1α CH$_3$, 7α CH$_3$	D-128	60	200-400	3-7	TP	SC
2α CH$_3$	D-45	20	400	20	MT	O
2α CHO	D-151	—	100	—	MT	O
2α SCOCH$_3$	D-44	6	6	—	MT	O
2α CH$_3$,17β Prop	D-6	50	200	4	TP	SC
2α CH$_3$, $\Delta^{9(11)}$	D-55	65	150	3	T	SC
$2\alpha,6\alpha$ CH$_3$, 17β Prop	D-7	50	200	4	TP	SC
2,2 (CH$_3$)$_2$	D-21	40	20	0.5	T	SC
2=CHOH	D-10	80	230	3	T	SC
2=CHOH	D-10	30-45	150-400	5-10	MT	O
2=CHOBenzoyl	D-16	25	20	0.8	T	SC
2=CH$_2$	D-152	15	15	3	MT	O
2=CHNH$_2$	D-78	20	160	8	MT	O
2=CHN(CH$_3$)$_2$	D-82	10	10	—	T	SC
2=CHN(CH$_3$)$_2$	D-82	20	100	5	MT	O
2=CHN(CH$_3$)ϕ	D-22	10	10	—	T	SC
2=CHN(cyclohexyl)	D-81	30	60	2	T	SC
2=CHN(cyclohexyl)	D-81	30	60	2	MT	O
2=CHN(Et)$_2$	D-83	20	120	6	MT	O
2 = CHNH(CH$_2$)$_2$N(CH$_3$)$_2$	D-79	30	150	5	MT	O
2 = CHNH(CH$_2$)$_2$N(Et)$_2$	D-80	40	65	1.6	T	SC
2 = CHNH(CH$_2$)$_2$N(Et)$_2$	D-80	40	100	2.5	MT	O
$4\alpha,5\alpha$ Epoxy, 11βOH	D-148	<40-60	<80-180	<2-3	MT	O
6α CH$_3$	D-13	30	50	1.6	T	SC
11β OH,11α CH$_3$	D-159	0	0	—	TP	SC

Table III.3

Compound	R	#	Andr	Anab	Q	St	Rte
Parent	Ac	D-99	30	100-200	3-6	TP	SC
1α CH$_3$	Ac	D-100	40	200-300	5-7	TP	SC
1α CH$_2$ Cl	Ac	D-104	2	10	—	TP	SC
1α CH$_2$ Br	Ac	D-105	4	30	7	TP	SC
1α CH$_2$ I	Ac	D-106	3	20	7	TP	SC
1α OH	Ac	D-107	70	200-300	3-4	TP	SC
1α OAc	Ac	D-108	70	200-400	3-6	TP	SC
1α CH(CH$_3$)$_2$	Ac	D-102	4.5	7.5	—	TP	SC
1β CH$_3$	Ac	D-101	1.5	10	—	TP	SC
$1{=}$CH$_2$	Ac	D-103	10	30	3	TP	SC
1 Oxo	Ac	D-109	13	100	7	TP	SC
1,1(CH$_3$)$_2$	Ac	D-110	4.5	7.5	—	TP	SC
1,1 ▽	Ac	D-111	4.5	7.5	—	TP	SC
1,1 O▽	Ac	D-115	4.5	7.5	—	TP	SC
$1\alpha,2\alpha$ —CH$_2$—	Ac	D-112	15	50-100	3-7	TP	SC
$1\alpha, 2\alpha$—CH$_2$— 4${=}$CHOH	Ac	D-113	4.5	7.5	—	TP	SC
$1\alpha, 2\alpha$— CH$_2$— 11β OH	Ac	D-114	4.5	7.5	—	TP	SC
$1\alpha,2\alpha$—CH$_2$— 2β Cl	Ac	D-118	4.5	7.5	—	TP	SC
$1\alpha,2\alpha$—CH$_2$— 2β Br	Ac	D-119	4.5	7.5	—	TP	SC
$1{=}$CH$_2$; 2,2(CH$_3$)$_2$	Ac	D-121	4.5	7.5	—	TP	SC
1αOH,2β Br	Ac	D-117	2	10-20	5-10	TP	SC
1αCH$_3$,2ξOAc	Ac	D-120	4.5	7.5	—	TP	SC
1αCH$_3$,7αCH$_3$	Ac	D-122	50-100	50-100	1	TP	SC
1αCH$_3$,7βCH$_3$	Ac	D-32	4.5	7.5	—	TP	SC
1ξ CH$_3$,2ξCH$_3$	Ac	D-116	15	25	1.7	TP	SC
2αCH$_3$	Prop	D-28	40	160	4	T	SC
2αCH$_3$	CH$_3$	D-19	10	20	2	T	SC
2αCH$_3$	COCHCl$_2$	D-158	75	206	3	T	SC

Table III. 3 (*continued*)

Compound	R	#	Andr	Anab	Q	St	Rte
2α CH$_3$	(ring with O and CH$_2$OH)	D-34	<100	<100	1	MT	O
2α CH$_3$	(ring with O)	D-35	70	307	4	MT	O
5α CH$_3$	Prop	D-143	0	0	—	T	SC
—	CH$_3$O (cyclopentyl)	D-130	196	308	1.5	MT	O
—	EtO (cyclopentyl)	D-132	200	300	1.5	MT	O

Table III.4

Compound	#	Andr	Anab	Q	St	Rte
Parent	D-145	22	18	0.8	T	SC
1α CH$_3$	D-129	<2	5	—	TP	SC
1α CHO	D-14	10	40	4	T	SC
1α,2α —CH$_2$—	D-131	4.5	20–30	5–7	TP	SC
1α CH$_3$, 3 Semicarb	D-176	23	60	2.6	T	SC
2 =CH$_2$	D-157	10	20	2	T	SC
2α,3α —CH$_2$—	D-40	30	100	3	TP	SC
2,2 O⟍	D-42	0	0	—	TP	SC
2α CH$_3$,3β OH	D-133	—	16	—	T	SC
2α CH$_3$,3α OH	D-164	—	25	—	T	SC
2α,3α Epoxy	D-64	1	1	—	TP	SC
2α,3α Epoxy	D-64	0	0	—	MT	O
2β,3β Epoxy	D-67	1.3	6.7	—	TP	SC
2β,3β Epoxy	D-67	0	0	—	MT	O
2α,3α Epithio	D-69	40	300	7	TP	SC
2α,3α Epithio	D-69	25	100	4	MT	O
2β,3β Epithio	D-65	2	13	6	TP	SC
2β,3β Epithio	D-65	0	0	—	MT	O
3α OH	D-31	25–280	30–65	0.3–1	T	SC
3β OH	D-146	5	—	—	T	SC
3,3 O⟍	D-43	10	10	—	TP	SC
3,3 O⟍	D-142	0	0	—	TP	SC
3α OH, 1α CH$_3$	D-171	105	164	1.5	T	SC
3α OH, 3β CH$_3$	D-163	0	—	—	T	SC
3β OH, 3α CH$_3$	D-162	0	—	—	T	SC
3β Cyclopentyl	D-156	5	5	—	T	SC
3 Fulvene	D-155	5	5	—	T	SC
3 Semicarbazone	D-175	60	76	1.2	T	SC
3 Isonicotinyl hydrazone	D-154	91	75	0.8	T	SC

Table III.5

Compound	R	#	Andr	Anab	Q	St	Rte
—	H	D-144	1.5	—	—	T	SC
17α CH$_3$	H	D-63	50	280	6	MT	O
—	OAc	D-134	1.5	3	—	TP	SC
1α CH$_3$	OAc	D-135	1.5	2.5	—	TP	SC
1β CH$_3$	OAc	D-136	1.5	2.5	—	TP	SC
1 Oxo	OAc	D-138	9	40	4	TP	SC
1α,2α —CH$_2$— 2=CH$_2$,	OAc	D-137	2	22	11	TP	SC
3β OAc	OAc	D-75	1	5	—	T	SC
2α,3α Epoxy	OAc	D-37	<10	<20	2	TP	SC
2β,3β Epoxy	OAc	D-36	10	10	—	TP	SC
2α,3α —CF$_2$—	OAc	D-51	35	130	3	T	SC
2β,3β —CF$_2$—	OAc	D-53	35	85	2	T	SC
3β OAc	OAc	D-29	13	8	—	T	SC
3α,4α Epoxy	OAc	D-39	<20	<20	1	TP	SC
3α OH	Oxo	D-27	50	30	0.6	T	SC
3β OH	Oxo	D-161	2	—	—	T	SC
3 Oxo	Oxo	D-30	20	10	0.5	T	SC

Table III.6

Compound	#	Andr	Anab	Q	St	Rte
Parent	D-12	100	183	2	MT	O
1α OH	D-59	5	23	4	TP	SC
1α OH	D-59	150	650	4	MT	O
1α CH$_3$	D-139	4.5	7.5	—	TP	SC
1β CH$_3$	D-140	1.5	2.5	—	TP	SC
1 Oxo	D-56	7	35	5	TP	SC
1 Oxo	D-56	180	830	4.5	MT	O
1α,2α —CH$_2$—	D-141	4	30	7	TP	SC
1α CH$_3$, 3 Oxime	D-174	70	170	2.5	T	SC
1α CH$_3$, [3,2-c] Pyrazole	D-177	30	60	2	T	SC
2β OH	D-60	0.5	1.5	—	TP	SC
2β OH	D-60	6	13	2	MT	O
2 Oxo	D-57	0.5	2	—	TP	SC
2 Oxo	D-57	4	16	4	MT	O
2 =CH$_2$	D-9	37	73	2	T	SC
2 =CH$_2$	D-9	100	200-400	2-4	MT	O
2α CH$_3$, 3 Hydrazone	D-15	100	280	3	MT	O
2 =CH$_2$, 3β OH	D-76	1	5	—	TP	SC
2 =CH$_2$, 3β OAc	D-77	1	5	—	TP	SC
2 =CH$_2$, 3β OAc	D-77	5	5	—	MT	O
(2α CH$_3$, 3 Azine) Dimer	D-5	8	24	3	T	SC
(2α CH$_3$, 3 Azine) Dimer	D-5	95	210	2	MT	O
2α,3α —CH$_2$—	D-41	<10	10	—	TP	SC
2α,3α —CF$_2$—	D-52	10	28	3	T	SC
2α,3α —CF$_2$—	D-52	50	250	5	MT	O
2α,3α Epoxy	D-38	1.5	5	—	TP	SC
2α,3α Epoxy	D-38	0	0	—	MT	O
2β,3β Epoxy	D-68	5	35	7	TP	SC
2β,3β Epoxy	D-68	0	0	—	MT	O
2α,3α Epithio	D-70	27	154	5	TP	SC
2α,3α Epithio	D-70	90	1100	12	MT	O
2β,3β Epithio	D-66	1	3	—	TP	SC
2β,3β Epithio	D-66	18	78	4	MT	O
2,3 Cyclohexeno	D-72	0	0	—	T	SC
2,3 Cyclohexano	D-73	0	0	—	T	SC
[3,2-b]Pyridine	D-149	2	5	—	TP	SC
[2,3-c]Oxadiazole	D-71	25	150	6	TP	SC
[2,3-c]Oxadiazole	D-71	80	300	4	MT	O
[3,2-c]Pyrazole	D-11	150	120	0.8	TP	SC
[3,2-c]Pyrazole	D-11	30	200	6	MT	O

Table III.6 (*continued*)

Compound	#	Andr	Anab	Q	St	Rte
[3,2-c]Isoxazole	D-33	20	150	7.5	MT	O
3α OH	D-61	4	1.6	—	TP	SC
3α OH	D-61	70	20	0.3	MT	O
3 =CH₂	D-54	60	69	1.1	MT	O
3,3 Azo	D-150	20	300	15	MT	O
3 Oxime	D-173	140	380	2.5	T	SC
3α OH, 11α CH₃, 11β OH	D-160	0	0	—	T	SC
4β OH	D-62	0.5	0.1	—	TP	SC
4β OH	D-62	20	35	2	MT	O
4 Oxo	D-58	0.6	1.8	—	TP	SC
4 Oxo	D-58	20	35	2	MT	O

209

Table IV

Number	Structure	Name	V.P.	S.V.	L.A.	Standard	Ref.	Adm.
A-1		5α-Androst-1-en-17β-ol	35 62	35 62	100 85	Testost Testost	43 82	17-Acetate
A-2		5α-Androst-2-en-17β-ol	0 37 50 — 5.8	0 42 50 15 5.8	0 134 150 30-40 17	17α-MT Testost Testost Test prop Test prop	87 398 43 100 89	Oral
A-3		5α-Androst-3-en-17β-ol	40 42 26	40 42 40	80 69 92	Testost Testost Testost	43 82 82	17-Acetate
A-4		Androst-4-en-17β-ol	10 20	10 20	10 18	Testost Testost	43 82	17-Acetate

Compound	Structure				Standard		Ref./Route
A-5	17α-Methyl androst-5-en-3β,17β-diol	38	29	18	Testost	107	
		—	5	7	Test prop	36	
		65	58	65	Test prop	32	
		—	68	80	17α-MT	42	Oral
		62	—	46	17α-MT	44	Oral
A-6	1-Methyl-5α-androst-1-en-17β-ol-3-one 17-acetate	15	—	72	17α-MT	44	Oral
		27	—	11.8	Test prop	81	(See also ref. 45)
		—	10	500	Test prop	47	
		—	20	100-200	Test prop	100	
		—	6	36	17α-MT	76	Oral
A-7	5α-Androst-1-en-17β-ol-3-one	—	100	200	Test prop	46	
		135	123	210	Testost	398	
		—	100	200	Test prop	85	
		—	5-10	30	Test prop	100	
A-8	17α-Methyl-17β-hydroxy-5α-androst-1-en-3-one	—	30	200	Test prop	100	
		—	25	50	Test prop	46	
		14	14	89	Test prop	84	
		180	220	910	17α-MT	84	Oral
		—	100	1600	17α-MT	85	Oral

Table IV (continued)

Number	Structure	Name	V.P.	S.V.	L.A.	Standard	Ref.	Adm.
A-9		17α-Ethyl-17β-hydroxy-5α-androst-1-en-one	—	2	5	Test prop	46	
A-10		2-Methyl-17β-hydroxy-5α-androst-1-en-3-one	— 130	50 115	50 300	Test prop Testost	46 82	
A-11		2,17α-Dimethyl-17β-hydroxy-5α-androst-1-en-3-one	— 17 90	25 17 170	50 56 660	Test prop Test prop 17α-MT	46 84 84	Oral
A-12		2-Methyl-17α-ethyl-17β-hydroxy-5α-androst-1-en-3-one	—	1	4	Test prop	46	

	Compound					
A-13	5α - Androst - 1 en - 3β,17β - diol	—	50	40	Test prop	46	
A-14	5α - Androst - 1 - en - 3β,17β - diol diacetate	—	50	40	Test prop	46	
A-15	17α - Methyl - 5α - androst - 1 - en - 3β,17β - diol	—	100	50	Test prop	46	
		15	13	24	Test prop	84	
		82	100	420	17α - MT	84	Oral
A-16	17α - Methyl - 5α - androst - en - 3β, 17β - diol 3 - monoacetate	—	100	200	Test prop	46	

Table IV (continued)

Number	Structure	Name	V. P.	S. V.	L. A.	Standard	Ref.	Adm.
A-17	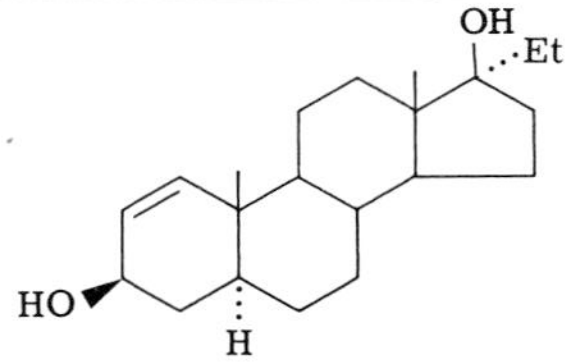	17α-Ethyl-5α-androst-1-en-3β,17β-diol	—	5	2	Test prop	46	
A-18	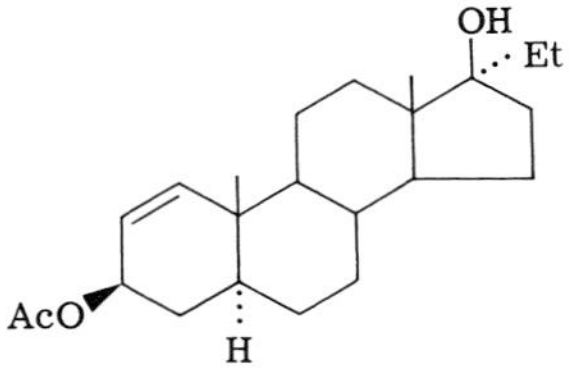	17α-Ethyl-5α-androst-1-en-3β,17β-diol-3-monoacetate	—	2	2	Test prop	46	
A-19	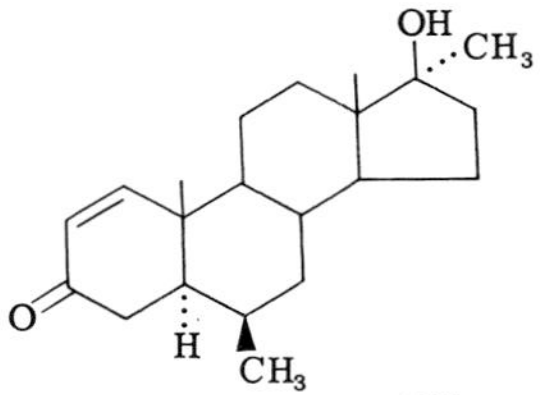	6β,17α-Dimethyl-17β-hydroxy-5α-androst-1-en-3-one	—	10	50	Test prop	46	
A-20	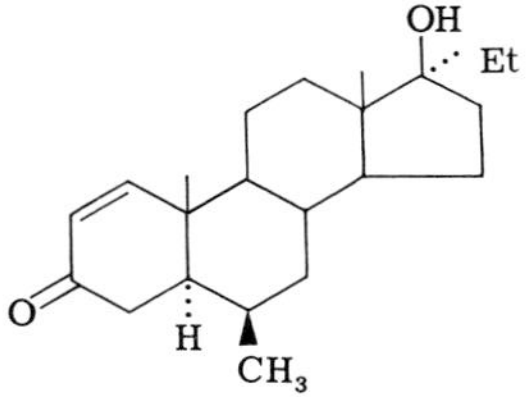	6β-Methyl-17α-ethyl-17β-hydroxy-5α-androst-1-en-3-one	—	<1	<5	Test prop	46	

214

No.	Compound				Standard		Route
A-21	3-Methylene-17α-methyl-5α-androst-1-en-17β-ol	—	12	266	17α-MT	48	Oral
A-22	17α-Methyl-17β-hydroxy-5α-androst-2-ene	187	187	1200	17α-MT	89	Oral
		4.8	4.8	11	Test prop	89	(See also ref. 100)
		70	88	362	Testost	30	
A-23	2-Methyl-17β-hydroxy-5α-androst-2-ene	119	218	437	17α-MT	82	Oral
		20	<10	100	Testost	56	
		50	50	200	17α-MT	43	Oral
		127	240	117	Testost	30	
		20	30	100	Testost	56	
		50	50	150	Testost	57	
A-24	2,17α-Dimethyl-17β-hydroxy-5α-androst-2-ene	64	87	310	Testost	30	Oral
		97	320	1040	17α-MT	30	Oral
		200	200	1000	17α-MT	57	

215

Table IV (continued)

Number	Structure	Name	V.P.	S.V.	L.A.	Standard	Ref.	Adm.
A-25		3,17α-Dimethyl-17β-hydroxy-5α-androst-2-ene	—	5.3	8	17α-MT	48	Oral
A-26		2-Hydroxymethyl-17α-methyl-17β-hydroxy-5α-androst-2-ene	45 19 30	62 44 30	203 218 220	Testost 17α-MT 17α-MT	30 30 58	Oral Oral
A-27		2-Cyano-17β-hydroxy-5α-androst-2-ene-17-caproate	125	170	870	Testost	30	
A-28		2-Formyl-17β-hydroxy-17α-methyl-5α-androst-2-ene	19 10 20 2.5 110	50 10 25 3.8 160	87 100 140 35 780	Testost Testost 17α-MT Test prop 17α-MT	30 56 82 84 84	Oral Oral

Code	Structure	Name				Standard		Route
A-29		2 - Difluoromethyl - 17β - hydroxy - 5α - androst - 2 - ene	4.4	8	7	Testost	30	Oral
			5	8	8	Testost	82	Oral
A-30		17α - Methyl - 17β - hydroxyandrostano- [2,3-d]isoxazole	24	—	200	17α - MT	11, 49	
			25	—	200	17α - MT	131	
			11	—	43	Test prop	11	
			65	81	134	Testost	398	
A-31		2,17α - Dimethyl - 17β - acetoxy - 5α - androst - 2 - ene	45	28	58	Testost	30	
A-32		3,17α - Dimethyl- androsta - 3,5 - dien - 17β - ol	—	31	25	17α - MT	48	Oral

217

Table IV *(continued)*

Number	Structure	Name	V. P.	S. V.	L. A.	Standard	Ref.	Adm.
A-33		17β - Hydroxy - 5α - androstano[3,2-b] quinoline	10	10	10	Testost	107	
A-34		4,4 - Dimethyl - 17β - hydroxyandrost - 5 - ene - 3 - one	0	0	0	Test prop	33	
A-35		2 - Methyl - 17β - acetoxy - 5α - androst - 1 - ene	20 28	20 24	50 52	Testost Testost	57 82	
A-36		2 - Formyl - 17β - hydroxy - 5α - androst - 2 - ene	20 10	20 10	100 100	Testost Testost	58 82	

A-37	OAc / OHC	2 Formyl - 17β - acetoxy - Δ²,⁴ - androstadiene	10	10	10	Testost	58
A-38	OH / HOCH₂	2 - Hydroxymethyl - 17β - hydroxy - 5α - androst - 2 - ene	20 / 66	20 / 50	220 / 457	Testost / Testost	58 / 82
A-39	OAc / ClCH₂	2 - Chloromethyl - 17β - acetoxy - 5α - androst - 2 - ene	10	10	10	Testost	59
A-40	OAc / F₂CH	2 - Difluoromethyl - 17β - acetoxy - 5α - androst - 2 - ene	10 / 52	10 / 48	10 / 156	Testost / Testost	59 / 82

219

Table IV (continued)

Number	Structure	Name	V.P.	S.V.	L.A.	Standard	Ref.	Adm.
A-41	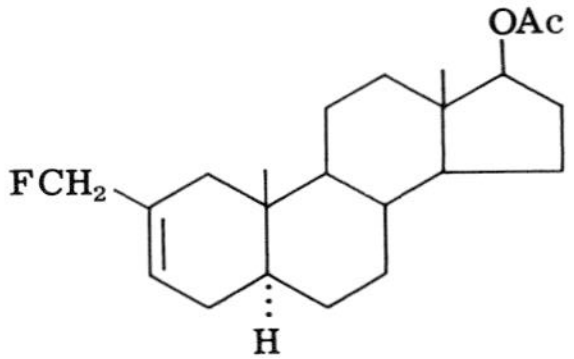	2 - Fluoromethyl - 17β - acetoxy - 5α - androst - 2 - ene	40	40	200	Testost	59	
A-42	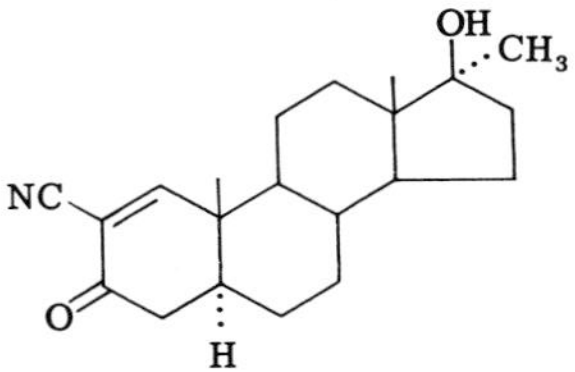	2 - Cyano - 17α - methyl - 17β - hydroxy - 5α - androst - 1 - en - 3 - one	10	<10	30	Testost	56	
A-43	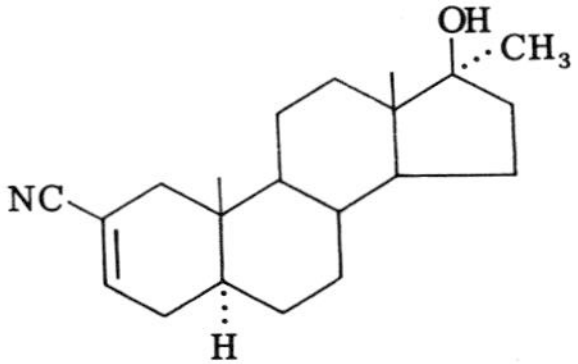	2 - Cyano - 17α - methyl - 17β - hydroxy - 5α - androst - 2 - ene	100 20	120 73	350 800	Testost 17α - MT	56 82	Oral
A-44	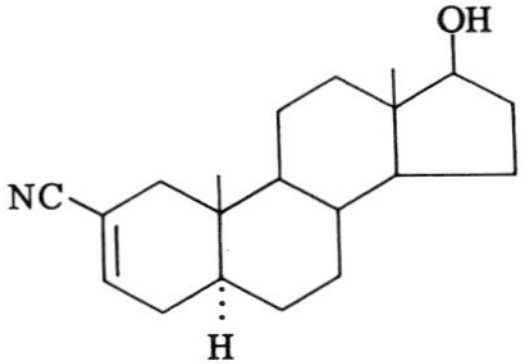	2 - Cyano - 17β - hydroxy - 5α - androst - 2 - ene	80 94	80 195	250 535	Testost Testost	56 82	

220

A-45		17β – Acetoxy – 5α- androsta – 1,3 – diene	25 44	30 44	100 93	Testost Testost	56 82
A-46		17β – Acetoxy -androsta – 2,4 – diene	40	50	100	Testost	56
A-47		2 - Acetoxymethylene - 17β – hydroxy – 5α– androst – 3 – ene	20	20	20	Testost	56
A-48		17α – Methyl – 17β – hydroxy – androsta – 3 – 5 – diene	5	5	20	Testost	56

221

Table IV (continued)

Number	Structure	Name	V. P.	S. V.	L. A.	Standard	Ref.	Adm.
A-49	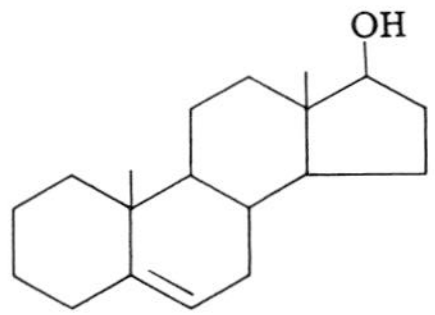	17β – Hydroxy-androst – 5 – ene	6	6	<10	Testost	56	
A-50	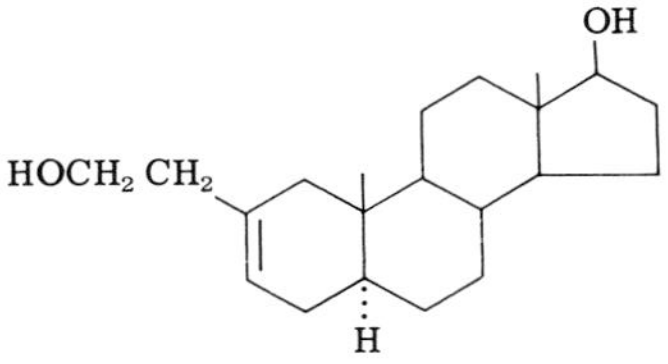	2 – Hydroxyethyl – 17β – hydroxy – 5α – androst – 2 – ene	<10 5	<10 5	<10 10	Testost Testost	56 82	
A-51	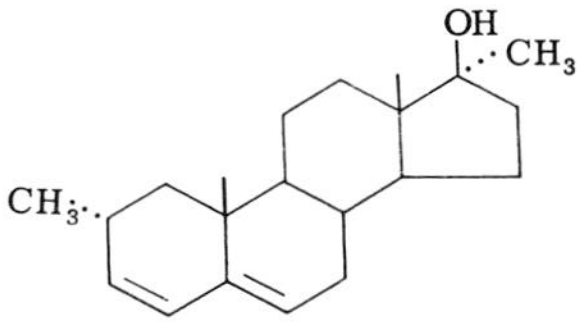	2α,17α – Dimethyl – 17β – hydroxy – androsta – 3,5 – diene	20	20	50	Testost	56	
A-52	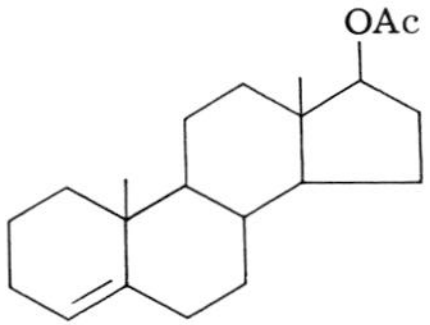	17β – Acetoxy – androst – 4 – ene	10	<10	<10	Testost	56	

222

No.	Structure	Name				Standard	Oral
A-53		2 - Cyano - 17α - methyl - 17β - acetoxy - 5α - androst - 2 - ene	<9 / 4	<9 / 17	<9 / 22	Testost / 17α - MT	56 / 82
A-54		3β,17β - Dihydroxy - androst - 4 - ene	124	133	95	Testost	61
A-55		3α,17β - Dihydroxy - androst - 4 - en 17β - acetate	112 / 74–112	155 / 87–155	94 / 90–94	Testost / Testost	61 / 107
A-56		3β,17β - Diacetoxy - androst - 4 - ene	13	—	6	Testost	61

Table IV (continued)

Number	Structure	Name	V.P.	S.V.	L.A.	Standard	Ref.	Adm.
A–57		3β – Hydroxyandrost-5 - en - 17 - one (Dehydroepi-androsterone)	<10 34	<10 14	<10 12	Testost Testost	62 107	Oral M.T.
A–58		17α - Methyl - 17β - hydroxy androst - 4 - eno[2,3-d] isoxazole	10 10–20 21	— — —	100 50–100 170*	17α – MT Test prop *Nitrogen retention	11,63 11 63	
A–59		17α- Methyl - 17β - hydroxy - androst - 5 - ene	—	116	95	17α – MT	42	Oral
A–60		2 - Methylthio - 17α- methyl - 17β - hydroxy - 5α- androst - 1 - en - 3 - one	—	8	11	17α – MT	76	Oral

No.	Compound						
A-61		2 – Methyl – 17β – acetoxy – 5α – androst – 1 – en – 3 – one	120 144	136 107	267 332	Testost Testost	82 82	
A-62		17α – Methyl – 17β – hydroxy – 5α – androst – 2 – ene – 2 – carboxylic acid	18	48	59	17α – MT	82	Oral
A-63		2 – Formyl – 17α – methyl – 17β – hydroxy – 5α – androst – 1 – en – 3 – one	60 50	53 50	93 250	17α – MT 17α – MT	82 155	Oral Oral
A-64		2 – Fluoromethyl – 17β – hydroxy – 5α – androst – 2 – ene	47	47	136	Testost	82	

Table IV (continued)

Number	Structure	Name	V. P.	S. V.	L. A.	Standard	Ref.	Adm.
A-65		4,4,17α-Trimethyl-17β-hydroxy-androst-5-en-3-one	0	0	0	Test prop	33	
A-66		17α-Methyl-17β-hydroxy-5α-androst-3-en-2-one	<0.2 16	<0.2 26	<0.4 <23	Test prop 17α-MT	84 84	Oral
A-67		17α-Methyl-17β-hydroxy-5α-androst-2-en-1-one	2 40 — —	2 35 10 20	9 140 20 150	Test prop 17α-MT Test prop 17α-MT	84 84 85 85	Oral Oral
A-68		17α-Methyl-17β-hydroxy-5α-androst-2-en-4-one	8.3 20 — —	8.3 29 20 15	34 170 100 100	Test prop 17α-MT Test prop 17α-MT	84 84 85 85	Oral Oral

No.	Structure	Name						
A-69		4,4,6,16α-Tetra-methylandrost-5-en-3β,17β-diol	10	14	0	Testost	95	
A-70		17α-Methyl-1α,17β-dihydroxy-5α-androst-2-ene	4	3	16	Test prop	84	Oral
			25	31	100	17α-MT	84	Oral
			—	10	100	17α-MT	85	
A-71		6β,17α-Dimethyl-17β-hydroxy-5α-androst-2-ene	<0.6	<0.6	<0.8	Test prop	84	Oral
			10	14	66	17α-MT	84	
A-72		1α,17α-Dimethyl-17β-hydroxy-5α-androst-2-ene	1.6	1.6	4.8	Test prop	84	Oral
			53	68	560	17α-MT	84	
			—	5	20	Test prop	100	

227

Table IV (continued)

Number	Structure	Name	V. P.	S. V.	L. A.	Standard	Ref.	Adm.
A-73		17α-Methyl-5α,17β-dihydroxy-androst-2-ene	0.7 <22	0.6 <23	1.3 <36	Test prop 17α-MT	84 84	Oral
A-74		17α-Methyl-2-bromo-17β-hydroxy-5α-androst-1-en-3-one	8.3 37	5.6 52 50	18 140 20-40	Test prop 17α-MT Test prop	84 84 29	Oral
A-75		2-Chloro-17α-methyl-17β-hydroxy-5α-androst-1-en-3-one	6.7 32 —	5.6 43 10	32 200 20	Test prop 17α-MT Test prop	84 84 29	Oral
A-76		2,17β-Dihydroxy-17α-methyl-5α-androst-1-en-3-one	<100	<140	<160	17α-MT	84	Oral

	Name					Oral
A-77	2 - Methoxy - 17α - methyl - 17β - hydroxy - androst - 1 - en - 3 - one	1.2 5.1	0.5 5.9	1.8 14	Test prop 17α - MT	84 84
A-78	17β - Acetoxy - 5α - androst - 1 - en - 3 - one	— —	100 50-100	400 100-200	Test prop Test prop	85 100
A-79	17β - Hydroxy - 5α - androst - 2 - en - 1 - one	—	1	5	Test prop	85
A-80	17β - Acetoxy - 5α - androst - 3 - en - 2 - one	—	1	4	Test prop	85

229

Table IV (continued)

Number	Structure	Name	V. P.	S. V.	L. A.	Standard	Ref.	Adm.
A-81		17β - Acetoxy - 5α - androst - 2 - en - 4 - one	—	25	200	Test prop	85	
A-82		4,4,6,16α - Tetra-methyl - 17β - hydroxyandrost - 5 - en - 3 - one	10	14	0	Testost	95	
A-83		17β - Hydroxy - 5α - androstano[3,2-e] 1′,4′ - thiaz - 5′ - en - 3′ - one	3	—	25	Test prop	92	
A-84		17β - Hydroxy - 17α - methyl - 5α - androstano[3,2-d] 2′ - methylpyrimidine	33 1.5	33 1.5	100 6	17α - MT Test prop	93 93	Oral

No.	Structure	Compound					Oral
A-85		17α-Methyl-17β-hydroxy-5α-androstano-2'-methyl[3,2-d]thiazole	10 / 40	10 / 40	30 / 200	Testost / 17α-MT	94 / 94
A-86		1-Methyl-17β-hydroxy-5α-androst-1-en-3-one	— / 44	1-2 / 57	30 / 88	Test prop / Testost	100 / 398
A-87		1-Ethyl-17β-hydroxy-5α-androst-1-en-3-one	— / 17	<4.5 / 11	<7.5 / 13	Test prop / Testost	100 / 398
A-88		1-Ethynyl-17β-hydroxy-5α-androst-1-en-3-one	—	<1.5	9	Test prop	100

Table IV (continued)

Number	Structure	Name	V.P.	S.V.	L.A.	Standard	Ref.	Adm.
A-89		1 - Methoxy - 17β - hydroxy - 5α - androst - 1 - en - 3 - one	—	<45	<45	Test prop	100	
A-90		2 - Chloro - 1 - methyl - 17β - hydroxy - 5α - androst - 1 - en - 3 - one	—	5	50	Test prop	100	
A-91		2 - Bromo - 1 - methyl - 17β - hydroxy - 5α - androst - 1 - en - 3 - one	—	2	10	Test prop	100	
A-92		2 - Methoxy - 1 - methyl - 17β - hydroxy - 5α - androst - 1 - en - 3 - one	—	<4.5	<7.5	Test prop	100	

A-93	OH, CH$_3$, H, CHOH	1 – Methyl – 4 – hydroxymethylene- 17β – hydroxy – 5α – androst – 1 – en – 3 – one	—	<4.5	< 7.5	Test prop	100
A-94	OH, HO, CH$_3$, H	1 – Methyl – 11β,17β – dihydroxy – 5α – androst – 1 – en – 3 – one	—	<4.5	10	Test prop	100
A-95	OH, HO, CH$_3$, H	1 – Methyl – 11α,17β- dihydroxy – 5α – androst – 1 – en – 3 – one	—	<4.5	<7.5	Test prop	100
A-96	OAc, Et, H	1 – Ethyl – 17β – acetoxy – 5α – androst – 1 – en – 3 – one	—	<4.5	<7.5	Test prop	100

233

Table IV (continued)

Number	Structure	Name	V.P.	S.V.	L.A.	Standard	Ref.	Adm.
A–97		1 - Chloro - 17β - acetoxy - 5α - androst - 1 - en - 3 - one	—	<4.5	<7.5	Test prop	100	
A–98		1 - Acetoxymethyl - 17β - acetoxy - 5α - androst - 1 - en - 3 - one	—	<4.5	<7.5	Test prop	100	
A–99		1 - Chloromethyl - 17β - acetoxy - 5α - androst - 1 - en - 3 - one	—	<1.5	4	Test prop	100	
A–100		1,2 - Dimethyl - 17β - acetoxy - 5α - androst - 1 - en - 3 - one	—	1	5	Test prop	100	

A-101	2 - Chloro - 1 - methyl - 17β - acetoxy - 5α - androst - 1 - en - 3 - one	—	5	50	Test prop 100
A-102	2 - Methoxy - 1 - methyl - 17β - acetoxy - 5α - androst - 1 - en - 3 - one	—	<1.5	<2.5	Test prop 100
A-103	1 - Methyl - 17β - acetoxy - 11β - hydroxy - 5α - androst - 1 - en - 3 - one	—	5	20-30	Test prop 100
A-104	1,17α - Dimethyl - 17β - hydroxy - 5α - androst - 1 - en - 3 - one	—	<15	200	Test prop 100

235

Table IV (*continued*)

Number	Structure	Name	V.P.	S.V.	L.A.	Standard	Ref.	Adm.
A-105		1,17α-Dimethyl-17β-acetoxy-5α-androst-1-en-3-one	—	<1.5	<2.5	Test prop	100	
A-106		1,17α-Diethyl-17β-hydroxy-5α-androst-1-en-3-one	—	<15	<25	Test prop	100	
A-107		1-Methyl-17α-vinyl-17β-hydroxy-5α-androst-1-en-3-one	—	<4.5	<7.5	Test prop	100	
A-108		1α-Methyl-17β-hydroxy-5α-androst-2-ene	— 34	3 28	30 81	Test prop Testost	100 398	

Compound	Name					
A-109	1β - Methyl - 17β - hydroxy - 5α - androst - 2 - ene	—	5	20–40	Test prop	100
A-110	1 - Methylene - 17β - hydroxy - 5α - androst - 2 - ene	—	6	20	Test prop	100
A-111	1,1 - Dimethyl - 17β - hydroxy - 5α - androst - 2 - ene	—	<4.5	9	Test prop	100
A-112	1,1 - Ethylene - 17β - hydroxy - 5α - androst - 2 - ene	—	4	20	Test prop	100

237

Table IV (*continued*)

Number	Structure	Name	V. P.	S. V.	L. A.	Standard	Ref.	Adm.
A-113	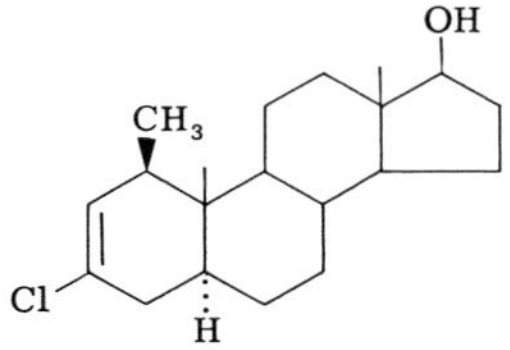	1 - Methyl - 3 - chloro - 17β - hydroxy - 5α - androst - 2 - ene	—	<1.5	<2.5	Test prop	100	
A-114	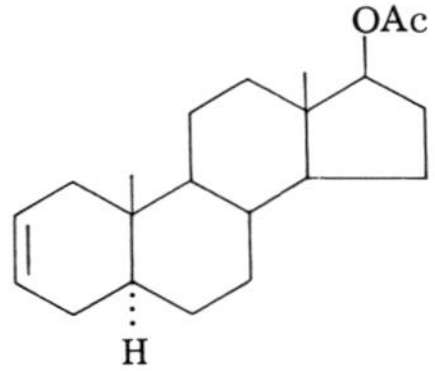	17β - Acetoxy - 5α - androst - 2 - ene	— 60	3 60	10–20 160	Test prop Testost	100 82	
A-115	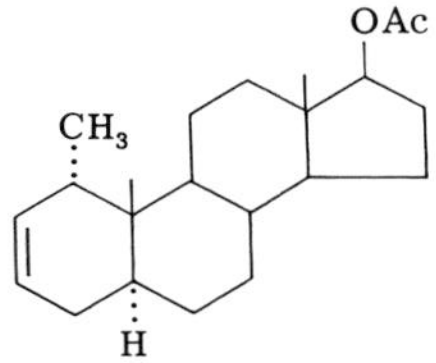	1α - Methyl - 17β - acetoxy - 5α - androst - 2 - ene	—	<1.5	6	Test prop	100	
A-116	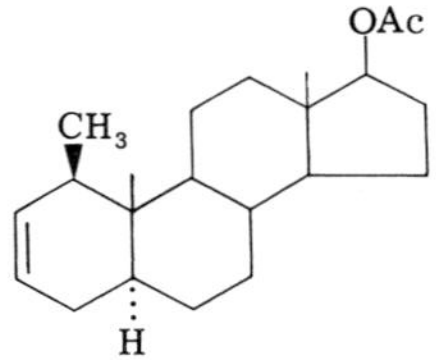	1β - Methyl - 17β - acetoxy - 5α - androst - 2 - ene	—	<1.5	7	Test prop	100	

238

Compound	Name					
A-117	1α - Hydroxy - 17β - acetoxy - 5α - androst - 1 - ene	—	<5	10-20	Test prop	100
A-118	1α - Chloro - 17β - acetoxy - 5α - androst - 1 - ene	—	10	40	Test prop	100
A-119	1α - Cyano - 17β - acetoxy - 5α - androst - 2 - ene	—	<1.5	<2.5	Test prop	100
A-120	1,1 - Dimethyl - 17β - acetoxy - 5α - androst - 2 - ene	—	<4.5	<7.5	Test prop	100

239

Table IV (continued)

Number	Structure	Name	V. P.	S. V.	L. A.	Standard	Ref.	Adm.
A-121	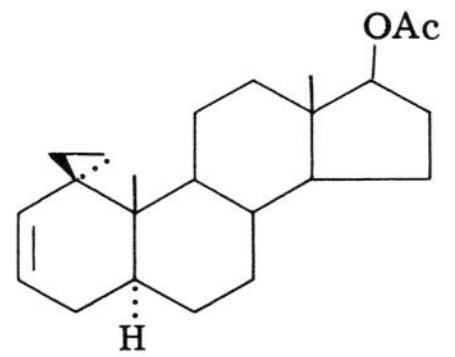	1,1 - Ethylene - 17β-acetoxy - 5α-androst - 2 - ene	—	<4.5	<7.5	Test prop	100	
A-122	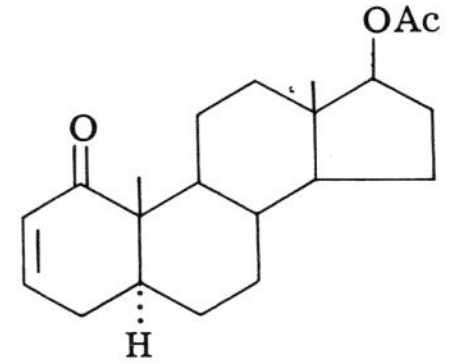	17β – Acetoxy - 5α-androst - 2 - en - 1 - one	—	<1.5	10	Test prop	100	
A-123	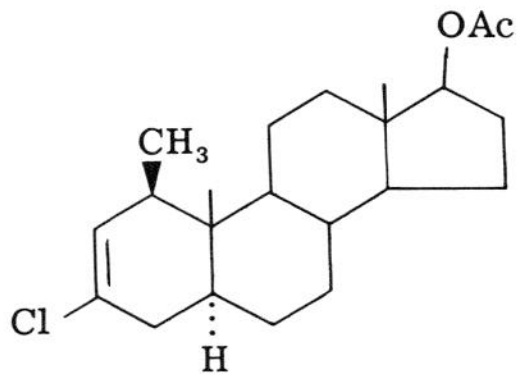	3 – Chloro - 1β-methyl - 17β-acetoxy - 5α-androst - 2 - ene	—	<4.5	<7.5	Test prop	100	
A-124	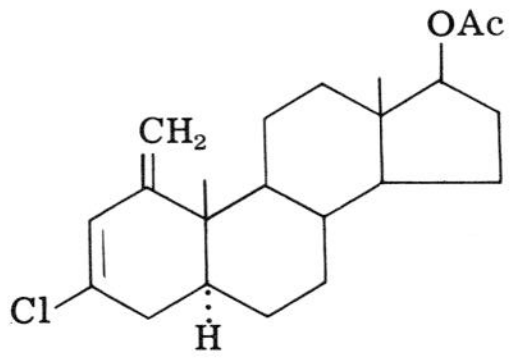	3 – Chloro - 1 - methylene - 17β-acetoxy - 5α-androst - 2 - ene	—	<1.5	<2.5	Test prop	100	

240

A-125	3 – Methyl – 17β – acetoxy – 5α – androst – 2 – en – 1 – one	—	<4.5	<7.5	Test prop	100
A-126	1α – Methyl – 17α – ethynyl – 17β – hydroxy – 5α – androst – 2 – ene	—	<1.5	<2.5	Test prop	100
A-127	1β,17α – Dimethyl – 17β – hydroxy – 5α – androst – 2 – ene	—	20	100	Test prop	100
A-128	1β – Methyl – 17α – ethyl – 5α androst – 2 – en – 17β – ol	—	<1.5	<2.5	Test prop	100

Table IV (continued)

Number	Structure	Name	V. P.	S. V.	L. A.	Standard	Ref.	Adm.
A-129		1β – Methyl – 17α – vinyl – 5α – androst – 2 – ene	—	1.5	2–3	Test prop	100	
A-130		1β – Methyl – 17α – ethynyl – 17β – hydroxy – 5α – androst – 2 – ene	—	<1.5	<2.5	Test prop	100	
A-131		Bis (1β – methyl – Δ² – 5α – androsten – 17β – ol – 17α – yl)	—	<1.5	<2.5	Test prop	100	
A-132		1 – Methylene – 17α – methyl – 17β – hydroxy – 5α – androst – 2 – ene	—	<5	10–20	Test prop	100	

	Compound				
A-133	1,1 - Ethylene - 17α - methyl - 17β - hydroxy - 5α - androst - 2 - ene	—	3	10	Test prop	100
A-134	1,1,17α - Trimethyl - 17β - hydroxy - 5α - androst - 2 - ene	—	<1.5	5	Test prop	100
A-135	3 - Chloro - 1β,17α - dimethyl - 5α - androst - 2 - en - 17β - ol	—	<1.5	<2.5	Test prop	100
A-136	1α - Methyl - 17β - hydroxyandrost - 4 - ene	—	<5	<8	Test prop	100

Table IV (*continued*)

Number	Structure	Name	V.P.	S.V.	L.A.	Standard	Ref.	Adm.
A-137		1 - Methyl - 17β - hydroxy - 5α - androst - 1 - ene	—	3-4	30	Test prop	100	
A-138		1 - Methyl - 17β - acetoxy - 5α - androst - 1 - ene	—	1.5	2-3	Test prop	100	
A-139		1α - Methyl - 17β - acetoxy - androst - 4 - ene	—	<1.5	<2.5	Test prop	100	
A-140		1α - Methyl - 17β - acetoxyandrosta - 4,6 - diene	—	<1.5	<2.5	Test prop	100	

					Test prop	100

A-141 — 1,17α-Dimethyl-17β-hydroxy-5α-androst-1-ene — | <45 | <45 | Test prop | 100

A-142 — 2-Chloro-17β-acetoxy-5α-androst-1-en-3-one — | 10 | 20 | Test prop | 29; — | 10 | 50 | Test prop | 379

A-143 — 2-Bromo-17β-hydroxy-5α-androst-1-en-3-one — | 10 | 10–20 | Test prop | 29

A-144 — 2-Chloro-17α-methyl-3β,17β-dihydroxy-5α-androst-1-ene — | 10 | 4–10 | Test prop | 29

Table IV (continued)

Number	Structure	Name	V.P.	S.V.	L.A.	Standard	Ref.	Adm.
A-145		2 - Chloro - 17α - methyl - 3β - acetoxy - 17β - hydroxy - 5α - androst - 1 - ene	—	5	4	Test prop	29	
A-146		17α - Methyl - 3β, 17β - dihydroxy - androst - 4 - ene	25–50 100	— —	— —	17α – MT Test prop	15 15	
A-147		17β - Acetoxy - Δ⁴ - androsteno[2,3-d] isoxazole	10	20	80	Testost	13	
A-148		17α - Methyl - 17β - hydroxyandrost - 4, 6 - dieno[2,3-d] isoxazole	1	—	2–5	Test prop	11	

A-149		17β - Hydroxy - androst - 4 - eno-[2,3-d]isoxazole	10-20 / 0	50-100 / 0	Test prop 11 / 17α- MT 11	Oral
A-150		17β - Hydroxy-androst - 3,5 - diene	⟨ Chick comb androgenic activity 10 ⟩		Testost 102
A-151		5α - Androst - 1 - en - 3,17 - dione	⟨ 9 By lacrymal gland ⟩		Testost 106
A-152		3β,17β - Dihydroxy-androst - 5 - ene	21	22	10	Testost 107

Table IV (continued)

Number	Structure	Name	V.P.	S.V.	L.A.	Standard	Ref.	Adm.
A-153		17β - Hydroxy-androstan - 2,4 - diene	40	40	120	Testost	155	
A-154		3α,17β - Dihydroxy - Δ⁴ - androstene 17 - (1' - ethoxy)cyclo-hexyl ether	—	150	120	17α - MT	12	Oral
A-155		3 - Chloro - 17β - acetoxyandrosta - 3,5 - diene	⟨ Favorable ratio ⟩				129	
A-156		17β - Hydroxy - 5α- androstano-[2,3-d]isoxazole	11 / 0	— / —	40 / 0	Test prop / 17α- MT	11 / 11	

Compound	Name						
A-157	17α - Methyl - 17β - hydroxyandrost - 4 - eno[3,2-c] pyrazole	26.5	—	47.5	D-11	124	Oral
		<3	—	<6	Test prop	124	Oral
		22	—	120	17α - MT	124	Oral
		46.5	—	61.5	D-11	124	Oral
A-158	17α - Methyl - 17β - hydroxy - 5α - androstano[3,2-d] thiazole	<25	<25	<25	17α - MT	94	Oral
A-159	17α - Methyl - 17β - hydroxy - 5α - androstano[3,2-d] 2' - phenylpyrimidine	20	20	<25	17α - MT	94	Oral
A-160	17α Methyl- testosterone 3 - cyclohexyl enol ether	210	385	270	17α - MT	139	Oral
		57	34	26	17α - MT	139	Oral
		500	500	500	17α - MT	140	Oral

Table IV (continued)

Number	Structure	Name	V.P.	S.V.	L.A.	Standard	Ref.	Adm.
A-161		Testosterone 3 - cyclopentenyl- enol ether	19 40	▽1 15	7 6	Test prop 17α- MT	141 141	Oral
A-162		17α - Methyl - testosterone - 3 - cyclopentyl enol ether.	38 165 225	11.5 295 420	23 250 285	Test prop 17α- MT 17α- MT	141 141 142	Oral Oral
A-163		4,5α - Dihydro, - Δ¹ - testosterone (1' - ethoxy) - 17 - cyclopentyl ether	—	250	760	17α- MT	12	Oral
A-164		4,5α - Dihydro - Δ¹ - testosterone 17 - cyclopent - 1' - enyl ether	—	240	400	17α- MT	12	Oral

	Compound					
A-165	17α- Propargyl-3β- acetoxy-androst - 5 - ene	0	0	—	Testost	163
A-166	3β,17β - Dihydroxy-3α- methylandrost-4 - ene	0	0	—	Test	3
A-167	3α,17β - Dihydroxy-3β - methyl-androst - 4 - ene	0	0	—	Test	3
A-168	17α- Methyl17β-hydroxyandrost-4,6 - dieno[3,2-c] pyrazole	0 0	— —	0 0	MT Stanozol (D-11)	124 124 Oral

Table IV (*continued*)

Number	Structure	Name	V. P.	S. V.	L. A.	Standard	Ref.	Adm.
A-169		17α - Methyl - 1,5 - androstadiene - 3β,17β - diol	31 60	— —	120 125	Stanozolol Δ¹-17α-MT	396 396	Oral Oral
A-170		1α,17α - Dimethyl - 17β - hydroxy - 5α - androstano[2,3-*d*] isoxazole	33	42	75	Testost	398	

Table IV.1

Compound	R	#	Andr	Anab	Q	St	Rte
—	—OH	A-7	10-100	30-200	2-3	TP	SC
—	—OAc	A-78	50-100	100-400	2-4	TP	SC
—	OH / CH$_3$	A-8	15-30	50-200	3-7	TP	SC
—	OH / CH$_3$	A-8	100-200	900-1600	5-16	MT	O
—	OH / Et	A-9	2	5	—	TP	SC
—	—O , EtO (spiro)	A-163	250	750	3	MT	O
—	—O (cyclopentene)	A-164	240	400	2	MT	O
—	Oxo	A-151	9	—	—	T	SC
1 CH$_3$	—OH	A-86	1-2	30	15	TP	SC
1 Et	—OH	A-87	<4.5	<7.5	—	TP	SC
1 C≡CH	—OH	A-88	<1.5	9	—	TP	SC
1 OCH$_3$	—OH	A-89	<45	<45	1	TP	SC
1 CH$_3$	—OAc	A-6	10-20	100-500	10-15	TP	SC
1 CH$_3$	—OAc	A-6	6	36	6	MT	O
1 Et	—OAc	A-96	<4.5	<7.5	—	TP	SC
1 Cl	—OAc	A-97	<4.5	<7.5	—	TP	SC
1 CH$_2$ OAc	—OAc	A-98	<4.5	<7.5	—	TP	SC
1 CH$_2$ Cl	—OAc	A-99	<1.5	4	—	TP	SC
1 CH$_3$	OH / CH$_3$	A-104	<15	200	13	TP	SC
1 CH$_3$	OAc / CH$_3$	A-105	<1.5	<2.5	—	TP	SC
1 Et	OH / Et	A-106	<15	<25	2	TP	SC
1 CH$_3$	OH / CH=CH$_2$	A-107	<4.5	<7.5	—	TP	SC
2 CH$_3$	—OH	A-10	50-130	50-300	1-2.5	TP, T	SC
2 Br	—OH	A-143	10	10-20	1-2	TP	SC

Table IV.1 (*continued*)

Compound	R	#	Andr	Anab	Q	St	Rte
2 CH$_3$	—OAc	A-61	100-140	260-330	3	T	SC
2 Cl	—OAc	A-142	10	20	2	TP	SC
2 CH$_3$	⟨OH / CH$_3$	A-11	20	50	2.5	TP	SC
2 CH$_3$	⟨OH / CH$_3$	A-11	130	660	4	MT	O
2 CH$_3$	⟨OH / Et	A-12	1	4	—	TP	SC
2 CN	⟨OH / CH$_3$	A-42	10	30	3	T	SC
2 SCH$_3$	⟨OH / CH$_3$	A-60	8	11	—	MT	O
2 CHO	⟨OH / CH$_3$	A-63	50	100-250	2-5	MT	O
2 Br	⟨OH / CH$_3$	A-74	8-50	20-40	0.8-2	TP	SC
2 Br	⟨OH / CH$_3$	A-74	40	140	3	MT	O
2 Cl	⟨OH / CH$_3$	A-75	5-10	20-30	2-5	TP	SC
2 Cl	⟨OH / CH$_3$	A-75	40	200	5	MT	O
2 OH	⟨OH / CH$_3$	A-76	120	160	1.2	MT	O
2 OCH$_3$	⟨OH / CH$_3$	A-77	5	14	3	MT	O
6 β CH$_3$	⟨OH / CH$_3$	A-19	10	50	5	TP	SC
6 β CH$_3$	⟨OH / Et	A-20	<1	<5	—	TP	SC
1 CH$_3$, 2 Cl	—OH	A-90	5	50	10	TP	SC
2 CH$_3$, 2 Br	—OH	A-91	2	10	5	TP	SC
1 CH$_3$, 2 OCH$_3$	—OH	A-92	<4.5	<7.5	—	TP	SC
1 CH$_3$, 4 =CHOH	—OH	A-93	<4.5	<7.5	—	TP	SC
1 CH$_3$, 11 β OH	—OH	A-94	<4.5	10	—	TP	SC
1 CH$_3$, 11 α OH	—OH	A-95	<4.5	<7.5	—	TP	SC
1 CH$_3$, 2 CH$_3$	—OAc	A-100	1	5	—	TP	SC
1 CH$_3$, 2 Cl	—OAc	A-101	5	50	10	TP	SC
1 CH$_3$, 2 OCH$_3$	—OAc	A-102	<1.5	<2.5	—	TP	SC
1 CH$_3$, 11 β OH	—OAc	A-103	5	20-30	4-6	TP	SC

Table IV.2

| Compound | | | | | | | | |
R$_2$		R$_1$	#	Andr	Anab	Q	St	Rte
—	3β OH	—OH	A-13	50	40	0.8	TP	SC
—	3β OH	⟨OH / CH$_3$⟩	A-15	15-100	25-50	0.5-2	TP	SC
—	3β OH	⟨OH / CH$_3$⟩	A-15	100	420	4	MT	O
—	3β OH	⟨OH / Et⟩	A-17	5	2	—	TP	SC
2 Cl	3β OH	⟨OH / CH$_3$⟩	A-144	10	5-10	—	TP	SC
—	3β OAc	—OAc	A-14	50	40	0.8	TP	SC
—	3β OAc	⟨OH / CH$_3$⟩	A-16	100	200	2	TP	SC
—	3β OAc	⟨OH / Et⟩	A-18	2	2	—	TP	SC
2 Cl	3β OAc	⟨OH / CH$_3$⟩	A-145	5	4	—	TP	SC
—	CH$_2$	⟨OH / CH$_3$⟩	A-21	10	60	6	MT	O
—	Δ^{3,4}	—OAc	A-45	25-44	95	2-4	T	SC
—	H$_2$	—OH	A-1	35	100	3	T	SC
1 CH$_3$ H$_2$		—OH	A-137	3-4	30	10	TP	SC
1 CH$_3$ H$_2$		—OAc	A-138	1.5	2-3	—	TP	SC
1 CH$_3$ H$_2$		⟨OH / CH$_3$⟩	A-141	<45	<45	1	TP	SC
—	H$_2$	—OAc	A-1	62	85	1.2	T	SC
2 CH$_3$ H$_2$		—OAc	A-35	25	50	2	T	SC
Δ^5	3β OH	⟨OH / CH$_3$⟩	A-169	60	125	2	Δ^1 MT	O

Table IV.3

Compound	R	#	Andr	Anab	Q	St	Rte
—	—OH	A-2	50	150	3	T	SC
—	—OAc	A-114	60	160	3	T	SC
—	—OH ⋯CH$_3$	A-22	20–80	100–360	5	T	SC
—	—OH ⋯CH$_3$	A-22	50–200	200–400	2–4	MT	O
1α CH$_3$	—OH	A-108	3	30	10	TP	SC
1α CH$_3$	—OAc	A-115	<1.5	6	—	TP	SC
1α CH$_3$	—OH ⋯CH$_3$	A-72	5	20	4	TP	SC
1α CH$_3$	—OH ⋯CH$_3$	A-72	60	560	9	MT	O
1α CH$_3$	—OH ⋯C≡CH	A-126	<1.5	<2.5	—	TP	SC
1α OH	—OAc	A-117	<5	10–20	4	TP	SC
1α OH	—OH ⋯CH$_3$	A-70	30	100	3	MT	O
1α Cl	—OAc	A-118	10	40	4	TP	SC
1α CN	—OAc	A-119	<1.5	<2.5	—	TP	SC
1β CH$_3$	—OH	A-109	5	20–40	4–8	TP	SC
1β CH$_3$	—OAc	A-116	<1.5	7	—	TP	SC
1β CH$_3$	—OH ⋯CH$_3$	A-127	20	100	5	TP	SC
1β CH$_3$	—OH ⋯Et	A-128	<1.5	<2.5	—	TP	SC
1β CH$_3$	—OH ⋯CH=CH$_2$	A-129	<1.5	2–3	—	TP	SC
1β CH$_3$	—OH ⋯C≡CH	A-130	<1.5	<2.5	—	TP	SC
1βCH$_3$ Dimer	—OH ⋯(2)	A-131	<1.5	<2.5	—	TP	SC
1,1(CH$_3$)$_2$	—OH	A-111	<4.5	9	—	TP	SC
1,1(CH$_3$)$_2$	—OAc	A-120	<4.5	<7.5	—	TP	SC
1,1(CH$_3$)$_2$	—OH ⋯CH$_3$	A-134	<1.5	5	—	TP	SC
1 Oxo	—OH	A-79	1	5	—	TP	SC
1 Oxo	—OAc	A-122	<1.5	10	—	TP	SC
1 Oxo	—OH ⋯CH$_3$	A-67	10	20	2	TP	SC
1 Oxo	—OH ⋯CH$_3$	A-67	30	150	5	MT	O
1 =CH$_2$	—OH	A-110	6	20	3	TP	SC
1 =CH$_2$	—OH ⋯CH$_3$	A-132	<5	10–20	4	TP	SC

Table IV.3 (*continued*)

Compound	R	#	Andr	Anab	Q	St	Rte
1,1 ▽	—OH	A-112	4	20	5	TP	SC
1,1 ▽	—OAc	A-121	<4.5	<7.5	—	TP	SC
1,1 ▽	⟍OH ⋯CH₃	A-133	3	10	3	TP	SC
1β CH₃, 3 Cl	—OH	A-113	<1.5	<2.5	—	TP	SC
1β CH₃, 3 Cl	—OAc	A-123	<4.5	<7.5	—	TP	SC
1β CH₃, 3 Cl	⟍OH ⋯CH₃	A-135	<1.5	<2.5	—	TP	SC
1 Oxo, 3 CH₃	—OAc	A-125	<4.5	<7.5	—	TP	SC
1 =CH₂, 3 Cl	—OAc	A-124	<1.5	<2.5	—	TP	SC
1α CH₃ [2,3-*d*] Isoxazole	⟍OH ⋯CH₃	A-170	35	75	2	T	SC
2 CH₃	—OH	A-23	30–240	100–150	0.5–4	TP	SC
2 CH₃	⟍OH ⋯CH₃	A-24	75	310	4	TP	SC
2 CH₃	⟍OAc ⋯CH₃	A-31	35	58	2	TP	SC
2 CH₃	⟍OH ⋯CH₃	A-24	100–300	1000	3–10	MT	O
2 CN	—OH	A-44	80–200	250–550	3–4	T	SC
2 CN	—OCapr	A-27	150	850	6	T	SC
2 CN	⟍OH ⋯CH₃	A-43	110	350	3	T	SC
2 CN	⟍CH ⋯CH₃	A-43	45	800	20	MT	O
2 CN	⟍OAc ⋯CH₃	A-53	<9	<9	—	T	SC
2 CN	⟍OAc ⋯CH₃	A-53	10	22	2	MT	O
2 CHO	—OH	A-36	15	100	6	T	SC
2 CHO	⟍OH ⋯CH₃	A-28	20–50	100	2–5	T	SC
2 CHO	⟍OH ⋯CH₃	A-28	20–110	140–780	7	MT	O
2 CHO, Δ⁴	—OAc	A-37	10	10	1	T	SC
2 CHF₂	—OH	A-29	8	8	—	T	SC
2 CHF₂	—OAc	A-40	50	150	3	T	SC
2 CH₂F	—OH	A-64	47	136	3	T	SC
2 CH₂F	—OAc	A-41	40	200	5	T	SC
2 CH₂OH	—OH	A-38	20–60	220–450	8–10	T	SC
2 CH₂OH	⟍OH ⋯CH₃	A-26	50	200	4	T	SC
2 CH₂OH	⟍OH ⋯CH₃	A-26	30	210	7	MT	O

Table IV. 3 (*continued*)

Compound	R	#	Andr	Anab	Q	St	Rte
$2CH_2Cl$	—OAc	A-39	10	10	—	T	SC
$2CH_2CH_2OH$	—OH	A-50	<10	<10	—	T	SC
2COOH	⟋OH ⋯CH_3	A-62	30	60	2	MT	O
$[2,3\text{-}d]$ isoxazole	—OH	A-156	11	40	4	TP	SC
$[2,3\text{-}d]$ isoxazole, Δ^4	—OH	A-149	10-20	50-100	5	TP	SC
$[2,3\text{-}d]$ isoxazole, Δ^4	⟋OH ⋯CH_3	A-58	10	100	10	TP / MT	SC / O
$[2,3\text{-}d]$ isoxazole, Δ^4	—OAc	A-147	20	80	4	T	SC
$[2,3\text{-}d]$ isoxazole Δ^4,Δ^6	⟋OH ⋯CH_3	A-148	1	2.5	—	TP	SC
$[2,3\text{-}d]$ isoxazole	⟋OH ⋯CH_3	A-30	25	200	8	MT	O
$[3,2\text{-}d]$ thiazole	⟋OH ⋯CH_3	A-158	<25	<25	1	MT	O
$[3,2\text{-}d]CH_3$ thiazole	⟋OH ⋯CH_3	A-85	40	200	5	MT	O
$[3,2\text{-}d]CH_3$ pyrimidine	⟋OH ⋯CH_3	A-84	33	100	3	MT	O
$[3,2\text{-}e]$ thiazenone	—OH	A-83	3	25	8	T	SC
$[3,2\text{-}b]$ quinoline	—OH	A-33	10	10	—	T	SC
$3\,CH_3$	⟋OH ⋯CH_3	A-25	5	8	—	MT	O
4 Oxo	—OAc	A-81	25	200	8	TP	SC
4 Oxo	⟋OH ⋯CH_3	A-68	20	100	5	TP	SC
4 Oxo	⟋OH ⋯CH_3	A-68	15-30	100-150	5-6	MT	O
Δ^4	—OH	A-153	40	120	3	T	SC
Δ^4	—OAc	A-46	45	100	2	T	SC
5α OH	⟋OH ⋯CH_3	A-73	<20	<35	1.5	MT	O
$6\beta\,CH_3$	⟋OH ⋯CH_3	A-71	12	66	5	MT	O

Table IV.4

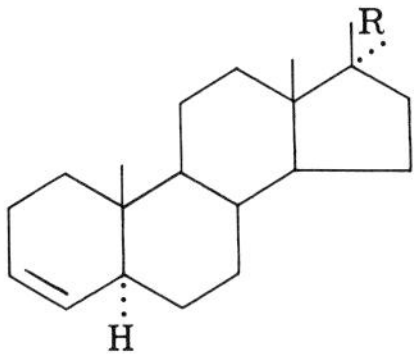

Compound	R	#	Andr	Anab	Q	St	Rte
—	—OH	A-3	40	80	2	T	SC
2=CHOAC	—OH	A-47	20	20	1	T	SC
2 Oxo	—OAc	A-80	1	4	—	TP	SC
2 Oxo	OH / CH₃	A-66	20	20	1	MT	O

Table IV.5

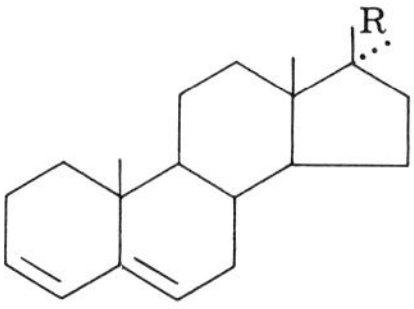

Compound	R	#	Andr	Anab	Q	St	Rte
—	—OH	A-150	10	—	—	T	SC
—	OH / CH₃	A-48	5	20	4	T	SC
2α CH₃	OH / CH₃	A-51	20	50	2.5	T	SC
3 CH₃	OH / CH₃	A-32	31	25	0.8	MT	O
3 Cl	—OAc	A-155	—	—	<1	T	SC
3 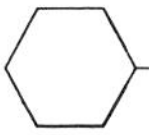 ⬡—O—	OH / CH₃	A-160	200-500	300-500	1	MT	O
3  ⬠—O—	—OH	A-161	30	6	0.2	MT	O
3 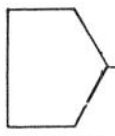 ⬠—O—	OH / CH₃	A-162	160-400	250-300	0.8	MT	O

259

Table IV.6

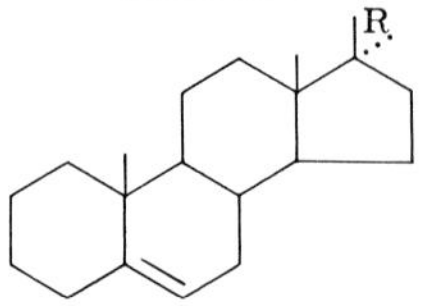

Compound	R	#	Andr	Anab	Q	St	Rte
—	−OH	A-49	6	<10	—	T	SC
—	OH ⁄ ··CH₃	A-59	116	95	0.8	MT	O
3β OH	−OH	A-152	20	10	0.5	T	SC
3β OH	OH ⁄ ··CH₃	A-5	30-60	20-60	1	TP	SC
3β OH	OH ⁄ ··CH₃	A-5	60	60	1	MT	O
3β OH	Oxo	A-57	<10	<10	—	T	SC
3β OAc	OH ⁄ ···Propargyl	A-165	0	0	—	T	SC
3βOH,4,4 (CH₃)₂ 6 CH₃, 16α CH₃	−OH	A-69	12	0	—	T	SC
3 Oxo, 4,4 (CH₃)₂	−OH	A-34	0	0	—	T	SC
3 Oxo,4,4 (CH₃)₂	OH ⁄ ···CH₃	A-65	0	0	—	T	SC
3 Oxo, 4,4 (CH₃)₂,.6 CH₃, 16α CH₃	−OH	A-82	10	0	—	T	SC

260

Table IV.7

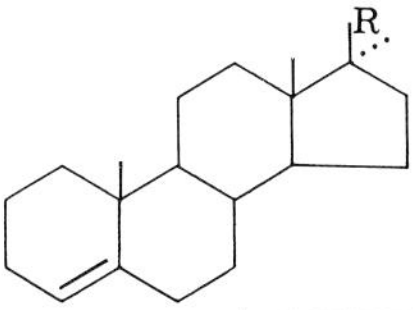

Compound							
R		#	Andr	Anab	Q	St	Rte
—	−OH	A-4	15	15	1	T	SC
—	−OAc	A-52	10	10	1	T	SC
1α CH$_3$	−OH	A-136	<5	<8	—	TP	SC
1α CH$_3$	−OAc	A-139	<1.5	<2.5	—	TP	SC
1α CH$_3$, Δ^6	−OAc	A-140	<1.5	<2.5	—	TP	SC
3α OH	−OAc	A-55	75-150	90	0.6-1.2	T	SC
3α OH	 ·O— / EtO	A-154	150	120	0.8	MT	O
3α OH, 3β CH$_3$	−OH	A-167	0	—	—	T	SC
3β OH	−OH	A-54	125	95	0.8	T	SC
3β OH	⁄OH ⋯CH$_3$	A-146	50-100	—	—	TP MT	SC O
3β OAc	−OAc	A-56	13	6	—	T	SC
3β OH, 3α CH$_3$	−OH	A-166	0	—	—	T	SC
[3,2-c] pyrazole	⁄OH ⋯CH$_3$	A-157	22	120	5	MT	O
[3,2-c] pyrazole, Δ^6	⁄OH ⋯CH$_3$	A-168	0	0	—	MT	O

261

Table V

Number	Structure	Name	V.P.	S.V.	L.A.	Standard	Ref.	Adm.
N-1	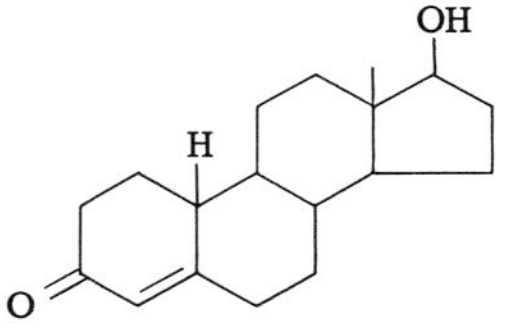	19-Nor-testosterone	—	7	66	Test prop	65	
			50	27	125	Testost	35	
			39	36	390	Testost	35	Oral
			—	6	100	Test prop	67	
			<10	20	200	Testost	64	
N-2	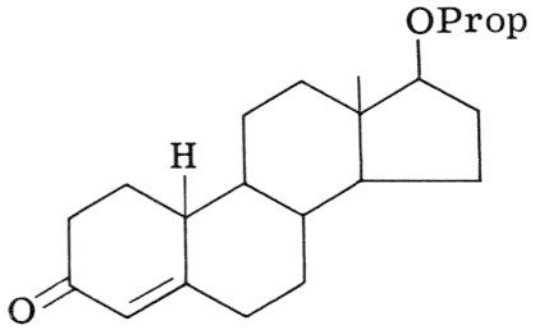	19-Nor-testosterone propionate	—	41	259	Test prop	65	[See also ref.105 (Lacrymal gland)]
N-3	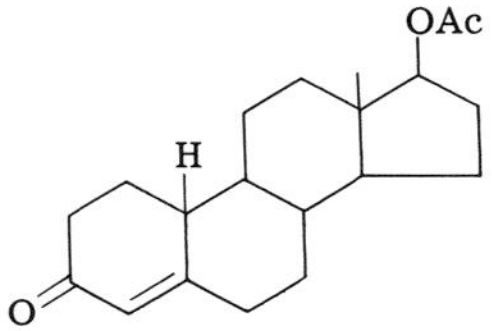	19-Nor-testosterone acetate	106	108	225	Testost	35	
			62	56	87	Testost	35	Oral
N-4	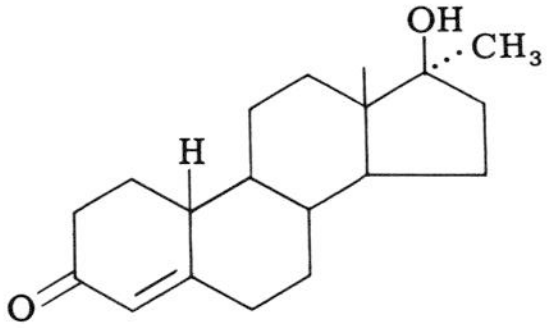	17α-Methyl-19-nortestosterone	—	6	100	Test prop	67	
			85	55	90	Testost	35	
			192	600	1420	Testost	35	
			37	37	450	Testost	64	Oral
			—	125	325	17α-MT	66	Oral
			110	—	580	17α-MT	44	Oral

262

	Name				Ref.		Route
N-5	17α-Ethyl-19-nortestosterone	38	38	77	Test prop	32	Oral
		—	33	103	17α-MT	42	
		12	—	80	Test prop	81	Oral
		—	7	104	17α-MT	36	
		57	24	200	Testost	67	
		37	29	120		79	
N-6	17α-Vinyl-19-nortestosterone	—	2	20	Test prop	67	
		18	7	24	Testost	79	
N-7	17α-Ethynyl-19-nortestosterone	3	10	40	Testost	64	
		—	1	1	Test prop	67	
		27	12	20	Testost	79	
		2	—	20	Test prop	91	
N-8	17α-Ethyl-17β-hydroxyestr-4-ene	—	22	421	17α-MT	42,280	Oral
		21	—	170	17α-MT	63	Oral
		17	—	40	17α-Ethyl-nortestost	77	
		40	40	200	17α-MT	107	Oral

Table V *(continued)*

Number	Structure	Name	V.P.	S.V.	L.A.	Standard	Ref.	Adm.
N-9		7α-Methyl-17β-acetoxy-estr-4-en-3-one	290	580	350	Testost	35	(See also ref. 105)
			165	145	165	Testost	35	Oral
			—	650	2300	Test prop	73	
			—	540	840	17α-MT	73	Oral
N-10		7α,17α-Dimethyl-17β-hydroxyestr-4-en-3-one	220	440	320	Testost	35	Oral
			320	540	365	Testost	35	Oral
			—	1800	4100	17α-MT	70	
			⟨519 By lacrymal gland⟩			19-Nortestost	105	
N-11		7α-Methyl-17β-hydroxy-estr-4-en-3-one	—	250	590	17α-MT	70	Oral
			—	300	1340	Test prop	70	
			⟨519 By lacrymal gland⟩			19 Nortestost	105	
N-12		4-Chloro-17α-methyl-17β-hydroxyestr-4-en-3-one	25	35	71	Test prop	32	

N-13		4 - Chloro - 17α-methyl - 17β-propionoxyestr - 4 - en - 3 - one	14	18	70	Test prop	60
N-14		4 - Chloro - 17β-acetoxyestr - 4 - en - 3 - one	26	20	112	Test prop	32
			3	3	77	Test prop	107
			—	40	660	Testost	126
N-15		4 - Methyl - 17β-hydroxyestr - 4 - en - 3 - one	5	5	20	Test prop	33
N-16		4 - Hydroxy - 17β-acetoxyestr - 4 - en - 3 - one	46	59	92	Test prop	32
			20	20	50	Test prop	107

265

Table V (continued)

Number	Structure	Name	V. P.	S. V.	L. A.	Standard	Ref.	Adm.
N–17		Estr - 4 - en - 3,17 - dione	45 55	16 31	58 37	Testost Testost	35 35	Oral
N–18		7α - Methylestr - 4 - en - 3,17 - dione	120 165	152 510	124 960	Testost Testost	35 35	Oral
N–19		3 - Methylene - 17α - methyl - 17β - hydroxyestr - 4 - ene	—	35	42	17α – MT	48	Oral
N–20		3 - Methylene - 17α - ethyl - 17β - hydroxyestr - 4 - ene	—	4	5	17α – MT	48	Oral

No.	Structure	Name					
N-21	CH₃ / OH / H / O steroid	16β - Methyl - 17β - hydroxy - estr - 4 - en - 3 - one	10	10	40	Test prop	60
N-22	OH / H / CH₃ / O steroid	2α - Methyl - 17β - hydroxy - 5α - estran - 3 - one	20	20	40	Test prop	60
			7	7	14	Test prop	107
N-23	OH···CH₃ / H / CH₃ CH₃ / O steroid	2,2,17α - Trimethyl - 17β - hydroxy - 5α - estran - 3 - one	3	5	2	Test prop	60
N-24	OH···CH₃ / H / CH₃ CH₃ / HO steroid	2,2,17α - Trimethyl - 3β,17β - dihydroxy - 5α - estrane	0	0	0	Test prop	60

Table V (continued)

Number	Structure	Name	V.P.	S.V.	L.A.	Standard	Ref.	Adm.
N–25		5ξ - Methyl - 17β - hydroxyestran- 3 - one	<5	<5	14	Testost	56	
N–26		17α - Methyl - 3β, 17β - dihydroxyestr - - 5 - ene	50	50	700	17α - MT	68	Oral
N–27		17α - Ethyl - 3β, 17β - dihydroxyestr - - 5 - ene	20	20	400	17α - MT	68	Oral
N–28		17β - Hydroxy - 5α- estr - 2 - ene	<10 2	<10 2	<10 20	Test prop Test prop	71 89	

		Oral
N-29	17α-Methyl-17β-hydroxy-5α-estr-2-ene	

Compound	Name				Standard	
N-29	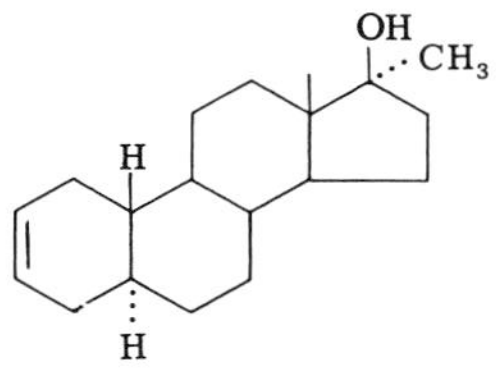 17α-Methyl-17β-hydroxy-5α-estr-2-ene	<10	<10	10	Test prop	71
		2	1.9	6.5	Test prop	84
		25	47	110	17α-MT	84
N-30	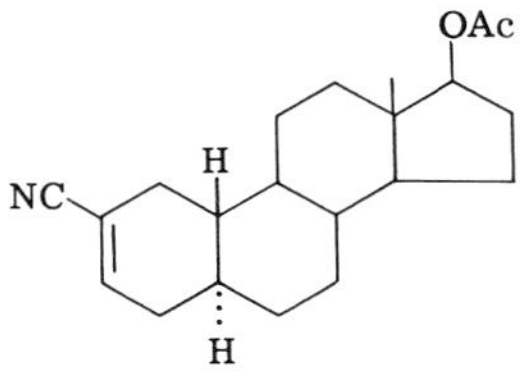 2-Cyano-17β-acetoxy-5α-estr-2-ene	30	50	100	Testost	56
N-31	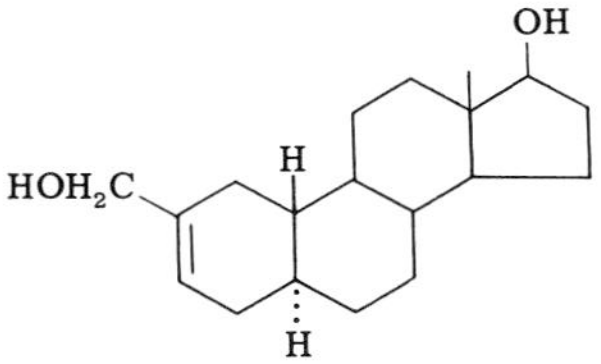 2-Hydroxymethyl-17β-hydroxy-5α-estr-2-ene	<15	<15	50	Testost	56
N-32	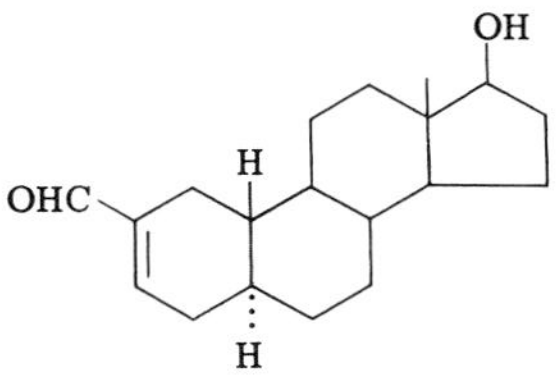 2-Formyl-17β-hydroxy-5α-estr-2-ene	<40	<10	30	Testost	56

OH CH₃ — N-29

OAc — N-30

OH, HOH₂C — N-31

OH, OHC — N-32

269

Table V *(continued)*

Number	Structure	Name	V.P.	S.V.	L.A.	Standard	Ref.	Adm.
N-33		17α - Allyl-estrenol	10 —	10 —	10 10	Testost Test prop	64 67	
N-34		3β,17β - Dihydroxy-estr - 4 - ene	—	5	50	Test prop	67	
N-35		17α - Propyl - 19 -nortestosterone	— 8	<1 7	10 11.5	Test prop Testost	67 79	
N-36		17α - Butyl - 19 -nortestosterone	—	<2	<5	Test prop	67	

No.	Name				Standard	Ref.
N-37	17α - Ethynyl - 17β - hydroxyestr - 5 - en - 3 - one	—	—	—	Same as N-7	68
N-38	17α - Ethyl - 17β - hydroxyestr - 5 - en - 3 - one	—	—	—	Same as N-5	68
N-39	17α - Methyl - 17β - hydroxyestr - 5 - en - 3 - one	—	—	—	Same as N-4	68
N-40	17α - Methyl - 17β - hydroxyestr - 5(10) - en - 3 - one	<5	<5	10	Test prop	69
		20	30	150	Testost	64
		—	—	5	Test prop	31

Table V (continued)

Number	Structure	Name	V.P.	S.V.	L.A.	Standard	Ref.	Adm.
N-41		4,4 – Dimethyl - 17β - hydroxyestr - 5 - en – 3 – one	0	0	0	Test prop	33	
N-42		4,4,17α– Trimethyl - 17β – hydroxyestr - 5 - en – 3 – one	0	0	0	Test prop	33	
N-43		17α– Ethynyl - 17β – hydroxyestr - 5(10) – en – 3 – one	<10	<10	<10	Testost	64	
N-44		2β,3β– Epoxy - 10 - cyano - 17β – acetoxy - 5α – estrane	0	0	0	Test prop	71	

Compound	Structure	Name			Reference compound	Ref.
N-45		13β – Ethyl - 17β - hydroxygon - 4 - en - 3 - one	625 27	— —	90 54	19 Nortestost Test prop	77 88,91
N-46		17α,13β – Diethyl - 17β - hydroxygon - 4 - en - 3 - one	280 17 —	— — —	860 350 1630	17α - Ethyl-nortestost Test prop MT	77 81,88,91 290
N-47		17α,13β – Diethyl - 17βhydroxygon - 4 - ene	120	—	160	17α - Ethyl-nortestost	77
N-48		1α - Methyl - 17β - hydroxy - 5α - estran - 3 - one	0	0	20	Testost	78

Table V (continued)

Number	Structure	Name	V.P.	S.V.	L.A.	Standard	Ref.	Adm.
N-49		17α-Methyl-17β,10-dihydroxy-5α-estr-2-ene	<0.6 9	<0.6 <7	1.3 <14	Test prop 17α-MT	84 84	Oral
N-50		17α-Methyl-17β-hydroxy-5α-estr-1-ene	1.6 110	2.3 200	2.4 170	Test prop 17α-MT	84 84	Oral
N-51		17α-Ethynyl-13β-ethyl-17β-hydroxygon-4-en-3-one	8 6	— —	70 50	Test prop Test prop	88 91	
N-52		13β-Ethyl-17β-hydroxy-7α-methylgon-4-en-3-one	60	—	300	Test prop	88	

No.	Structure	Name				Ref.	
N-53		13β,17α-Diethyl-17β-hydroxy-7α-methylgon-4-en-3-one	70	—	48	Test prop	88
N-54		13β-Ethyl-17β-ethynyl-17β-hydroxy-7α-methylgon-4-en-3-one	26	—	27	Test prop	88
N-55		17β-Hydroxy-10-hydroxymethyl-5α-estr-2-ene	0.2 / 0	0.2 / 0	0.6 / 0	Test prop / 17α-MT	89 / 89	Oral
N-56		17α-Methyl-17β-hydroxy-10-hydroxymethyl-5α-estr-2-ene	<0.6 / 0	<0.6 / 0	1.3 / 0	Testost / 17α-MT	89 / 89	Oral

Table V *(continued)*

Number	Structure	Name	V.P.	S.V.	L.A.	Standard	Ref.	Adm.
N-57		10 - Cyano - 3β, 17β - dihydroxyestr - 5 - ene	0	0	0	Test prop	99	
N-58		10 - Cyanoestr - 4 - en - 3,17 - dione	0	0	0	Test prop	99	
N-59		10 - Formyl - 17α- methyl - 3β,17β - dihydroxyestr - 5 - ene	<10	<10	<10	Test prop	99	
N-60		10 - Hydroxymethyl - 17α- methyl - 3β, 17β - dihydroxyestr - 5 - ene	< 10	<10	0	Test prop	99	

276

N-61	1α - Methyl - 17β - acetoxy - 5α - estran - 3 - one	—	<1.5	20-30	Test prop	100
N-62	1α - Hydroxy - 17β - acetoxy - 5α - estran - 3 - one	—	2-3	20-30	Test prop	100
N-63	1,2α - Methylene - 17β - acetoxy - 5α - estran - 3 - one	—	4.5	20-30	Test prop	100
N-64	1 - Methyl - 17β - acetoxy - 5α - estr - 1 - en - 3 - one	—	<4.5	<4.5	Test prop	100

Table V *(continued)*

Number	Structure	Name	V.P.	S.V.	L.A.	Standard	Ref.	Adm.
N-65	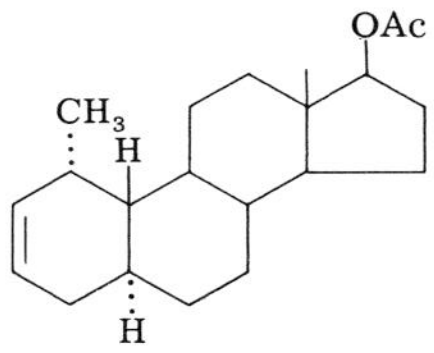	1α - Methyl - 17β - acetoxy - 5α - estr - 2 - ene	—	< 1.5	< 2.5	Test prop	100	
N-66	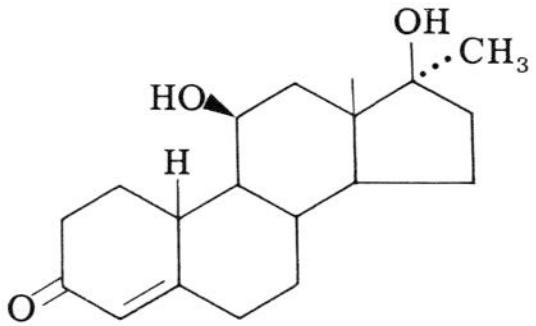	17α - Methyl - 11β, 17β - dihydroxyestr - 4 - en - 3 - one	—	260	800	17α - MT	24	
N-67	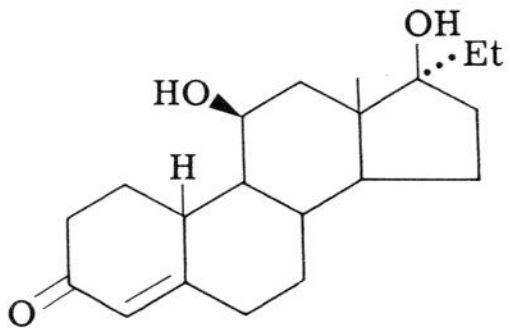	17α - Ethyl - 11β, 17β - dihydroxy- estr - 4 - en - 3 - one	—	80	110	17α - MT	24	
N-68	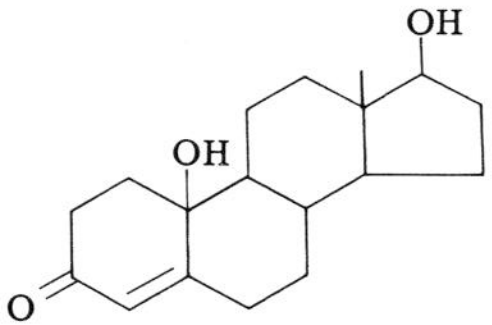	10β,17β - Di- hydroxyestr - 4 - en - 3 - one	—	〈Ratio given, 1.5〉 —	< 20	Testost Test prop	14 31	17 - Acetate

278

N-69		5α - Fluoro - 10β, 17β - dihydroxy - 17α - ethynyl- estran - 3 - one	⟨Ratio given, 1.7⟩		Testost	14
N-70		17β - Acetoxy - Δ⁴ - estreno[2,3 - d] isoxazole	10	20	80	Testost	13
N-71		Estr - 5(10) - ene - 17β - ol 17 - cyclopent - 1' - enyl ether	—	40	120	17α - MT	12	Oral
N-72		17α - Methyl - 17β - hydroxy - 19 - norandro- stano [2,3-d] isoxazole	10-20	—	10-20	Test prop	11

Table V *(continued)*

Number	Structure	Name	V.P.	S.V.	L.A.	Standard	Ref.	Adm.
N-73		7α - Methyl - 17α - ethynyl - 17β - hydroxy - estr - 4 - en - 3 - one	130	105	—	17α - MT	103	Oral
N-74		4 - Hydroxy - 17α - methyl - 17β - hydroxyestr - 4 - en - 3 - one	1024 71 281	— — —	1304 105 1304	17α - Methyl - nortestost 17α - MT	127 127 127	Oral Oral
N-75		19 - Nor - testosterone dimethyl - hydrazone	5	<1	10	Test prop	128	Oral
N-76		17α - Methyl - 17β - hydroxyestr - 4 - ene[2,3 - *d*] isoxazole	20-40	—	100	Test prop	130	Oral

No.	Structure	Name				Relative to testost 17-cyclopentyl-propionate
N-77	OCOC$_2$H$_4$- (cyclopentyl) steroid, 4-OH, 3-keto	4-Hydroxy-19-nortestosterone 17-cyclopentyl-propionate	40 32 33 25	1st week 2nd week 3rd week 4th week	120 100 160 150	244
N-78	OCOC(CH$_3$)$_3$ steroid	Nortestosterone 17-trimethyl-acetate	—	< 2	18	Test prop 65
N-79	OCOC$_6$H$_5$ steroid	Nortestosterone 17-benzoate	—	2.5-5.0	50	Test prop 107
N-80	OCOC$_2$H$_4$- (cyclopentyl) steroid	Nortestosterone 17-cyclopentyl-propionate (ester)	—	20	202	Test prop 65

Table V (*continued*)

Number	Structure	Name	V. P.	S. V.	L. A.	Standard	Ref.	Adm.
N-81		Nortestosterone 17 - cyclopen - 1' - enyl ether	—	40	440	17α - MT	12	Oral
N-82		17α - Methyl - nortestosterone - 17 - cyclopent - 1' - enyl ether	—	145	364	17α - MT	107	
N-83		17α - Methyl - 17β - hydroxy - estr - 5 (10) - ene - 17 - cyclopent - 1' - enyl ether	—	130	286	17α - MT	107	
N-84		19 - Nortestosterone 17 - adamantoate [same dose (once)]	1st week 3rd week 6th week 8th week	5.5 11.2 8.8 9.4	26.4 58.1 103.6 97.3	Relative s.c.* to N-85	147	

282

	Structure	Compound	Week				Ref.
N-85	OCOC$_9$H$_{19}$-n	19-Nortestosterone 17-n-decanoate [same dose (once)]	1st week 3rd week 6th week 8th week	55.7 57.5 104.2 72.6	45.4 98.8 134.9 83.7	Relative s.c.*	147
N-86	OCOC$_2$H$_4$Ph	19-Nortestosterone 17-phenylpropionate [same dose (once)]	1st week 2nd week 4th week 6th week	23.8 13.9 15.8 13.1	37.4 32.0 53.3 60.4	Relative s.c.* to N-87	148
N-87	OCOC$_9$H$_{19}$-n	19-Nortestosterone 17-n-decanoate [same dose (once)]	1st week 2nd week 4th week 6th week	19.7 13.3 17.7 13.8	46.3 60.9 110.2 68.8	Relative s.c.*	148
N-88	OH ... CH$_3$	6α-Methyl-17β-hydroxy-estr-4-en-3-one		17	17	26 Test prop	107

*s. c. = subcutaneous administration

283

Table V *(continued)*

Number	Structure	Name	V.P.	S.V.	L.A.	Standard	Ref.	Adm.
N-89		17β - Hydroxy - 5α - estran - 3 - one	—	—	<10	Test prop	31	
N-90		17α - Methyl - 17β - hydroxy - 5α - estran - 3 - one	10	10	100	Test prop	31	
N-91		17α - Ethyl - 17β - hydroxy - 5α - estran - 3 - one	—	—	10	Test prop	31	
N-92		17β - Hydroxyestr - 5(10) - en - 3 - one	—	—	5	Test prop	31	

No.	Structure					Standard	
N-93	17α - Ethyl - 17β - hydroxy - estr - 5(10) - en - 3 - one	—	—	5	Test prop	31
N-94	17α - Ethyl - 3β, 17β - dihydroxy - estr - 4 - ene	—	—	50	Test prop	107
N-95	17β - Hydroxy - estra - 4,9(10) - dien - 3 - one	10	10	100	17α - MT	154
N-96	17α - Methyl - 17β - hydroxy - estra - 4,9(10) - dien - 3 - one	30	30	130	17α - MT	
		200	300	1000	17α - MT	Oral

Table V (continued)

Number	Structure	Name	V.P.	S.V.	L.A.	Standard	Ref.	Adm.
N-97		17α-Ethyl-17β-hydroxyestra-4,9(10)-dien-3-one	10 30 130	10 30 —	100 150 <60	17α-MT 17α-MT 17α-Ethylnor-testost	154 154 77	Oral
N-98		13β,17α-Diethyl-17β-hydroxygona-4,9(10)-dien-3-one	860	—	1450	17α-Ethylnor-testost	77	
N-99		17α-Methyl-17β-hydroxyestra-4,9,11-triene-3-one	6000 —	7500 —	12,000 30,000	17α-MT 17α-MT	55 74,75	Oral
N-100		10-Vinyl-19β-hydroxyestr-4-en-3-one	75	89	52	Testost	82	

	Structure	Name				Reference		
N-101		10 - Ethyl - 19β - hydroxyestr - 4 - en - 3 - one	5	3	4	Testost	82	Oral
N-102		7α,17α - Dimethyl - 17β - hydroxyestra - 4,9,11 - triene - 3 - one	over 10,000	over 10,000	over 10,000	17α - MT	159	Oral
N-103		17β - Acetoxyestra - 4,9,11 - trien - 3 - one	500	500	500	Nortestost acetate	159	
N-104		17β - Methoxy- methyloxyestra - 4,9,11 - trien - 3 - one	2000	2000	2000	17α - MT	159	Oral

Table V *(continued)*

Number	Structure	Name	V.P.	S.V.	L.A.	Standard	Ref.	Adm.
N-105	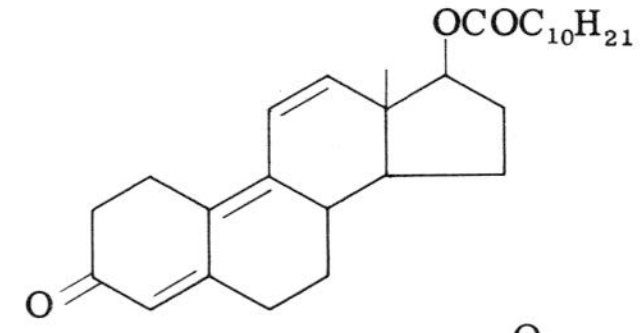 $OCOC_{10}H_{21}$	17β-Hydroxyestra-4,9,11-trien-3-on-17-n-undecanoate	500	500	500	Nortestost β-phenylpro-pionate	159	
N-106	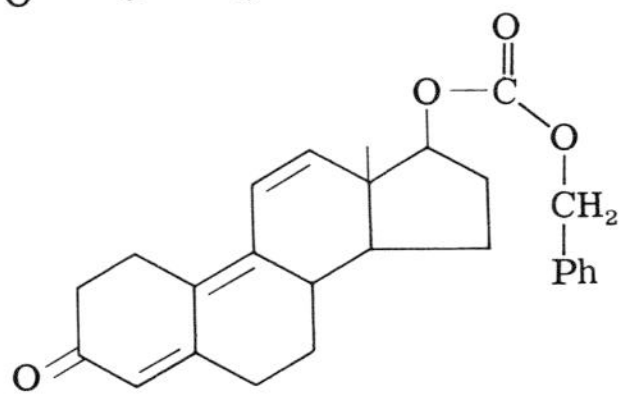	17β-Hydroxyestra-4,9,11-trien-3,one-17-carbo-benzoxyate	500	500	500	Nortestost β-phenylpro-pionate	159	
N-107	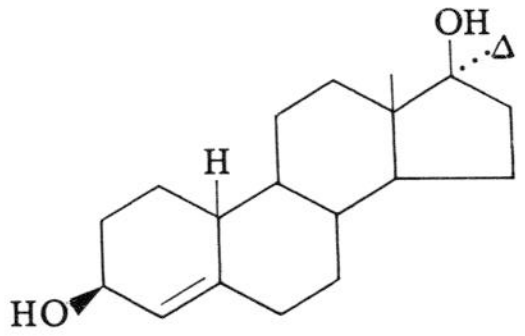	17α-Cyclopropyl-17β,3β-hydroxy-4-estrene	6	6	50	17α-MT	392	Oral

288

Table V.1

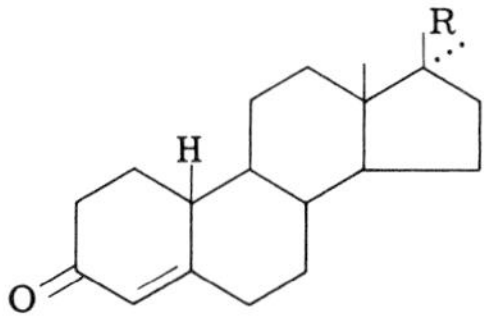

Compound	R	#	Andr	Anab	Q	St	Rte
—	—OH	N-1	37	125	4	T	SC
—	—OH	N-1	37	390	10	T	O
—	—OProp	N-2	40	260	6	TP	SC
—	—OAc	N-3	60-100	90-220	1.5-2	T	SC
—	—OH ···CH$_3$	N-4	30-80	100-450	2-10	T	SC
—	—OH ···CH$_3$	N-4	100	300-600	3-6	MT	O
—	—OH ···Et	N-5	40	120	3	T	SC
—	—OH ···Et	N-5	25-55	100-200	3-4	MT	O
—	—OH ···CH=CH$_2$	N-6	10	20	2	T	SC
—	—OH ···C≡CH	N-7	10	20	2	T	SC
—	—OH ···C$_3$H$_7$	N-35	7	10	1.3	T	SC
—	—OH ···C$_4$H$_9$	N-36	<2	<5	—	TP	SC
—	—O— 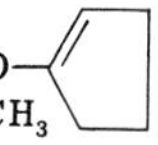 ···CH$_3$ (cyclopentenyl)	N-82	145	364	2	MT	O
—	=O	N-17	30	50	1.5	T	SC
—	—OCOC(CH$_3$)$_3$	N-78	<2	18	9	TP	SC
—	—OCOCC$_6$H$_5$	N-79	5	50	10	TP	SC
—	OCO—C$_2$H$_4$—(cyclopentyl) 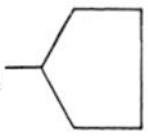	N-80	20	200	10	TP	SC
—	—O—(cyclopentenyl) 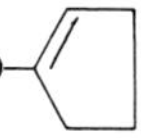	N-81	40	440	11	MT	O
—	—O— Adamant.	N-84	—	—	10	—	SC

289

Table V.1 *(continued)*

Compound	R	#	Andr	Anab	Q	St	Rte
—	—O— Decanoate	N-85	—	—	2	—	SC
—	—O— Decanoate	N-87	—	—	6	—	SC
—	OCO | CH$_2$C$_6$H$_5$	N-86	—	—	4	—	SC
3 Dimethyl-hydrazone	OH	N-75	5	10	2	TP	O
2,2(CH$_3$)$_2$	OH / CH$_3$	N-23	4	2	—	T	SC
4 Cl	OH / CH$_3$	N-12	30	71	2	TP	SC
4 Cl	OProp / CH$_3$	N-13	16	70	4	TP	SC
4 Cl	—OAc	N-14	22	112	5	TP	SC
4 CH$_3$	—OH	N-15	5	20	4	TP	SC
4 OH	—OAc	N-16	20-60	50-90	1.5-2	TP	SC
4 OH	OCO C$_2$H$_4$—	N-77	35	130	3.7	TCPP	SC
4 OH	OH / CH$_3$	N-74	280	1300	4.5	MT	O
6α CH$_3$	—OH	N-88	17	26	1.5	TP	SC
7α CH$_3$	—OH	N-11	300	1350	4.5	TP	SC
7α CH$_3$	—OH	N-11	250	590	2	MT	O
7α CH$_3$	=O	N-18	300	900	3	T	O
7α CH$_3$	=O	N-18	120	120	1	T	SC
7α CH$_3$	OH / CH$_3$	N-10	350	350	1	T	SC
7α CH$_3$	OH / CH$_3$	N-10	1800	4100	2.5	MT	O
7α CH$_3$	OH / C≡CH	N-73	115	—	—	MT	O
7α CH$_3$	—OAc	N-9	300-600	350-2300	2-4	T	SC
7α CH$_3$	—OAc	N-9	540	840	1.6	MT	O
11β OH	OH / CH$_3$	N-66	260	800	3	MT	O
11β OH	OH / Et	N-67	80	110	1	MT	O
16β OH	—OH	N-21	10	40	4	TP	SC

Table V.2

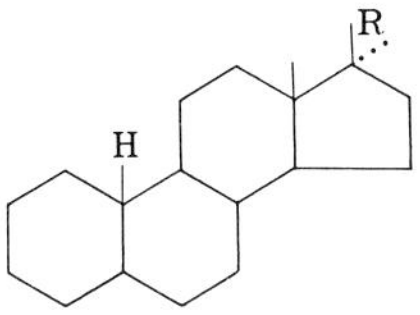

Compound	R	#	Andr	Anab	Q	St	Rte
Δ^1	OH, CH$_3$	N-50	150	170	1.2	MT	O
Δ^1, 1CH$_3$	—OAc	N-64	<4.5	<4.5	—	TP	SC
Δ^2	—OH	N-28	2	20	10	TP	SC
Δ^2	OH, CH$_3$	N-29	35	110	3	MT	O
Δ^2, 2CH$_2$ OH	—OH	N-31	15	50	3	T	SC
Δ^2, 2CHO	—OH	N-32	25	30	1.2	T	SC
Δ^2, 1αCH$_3$	—OAc	N-65	<4.5	<4.5	—	TP	SC
Δ^2, 2CN	—OAc	N-30	40	100	2.5	T	SC
Δ^4	OH, Et	N-8	20-40	200-400	5-20	MT	O
Δ^4	OH, Allyl	N-33	10	10	—	T	SC
Δ^4, 3β OH	OH, Et	N-94	—	50	—	TP	SC
Δ^4, 3β OH	OH, Δ	N-107	6	50	8	MT	O
Δ^4, 3β OH	—OH	N-34	5	50	10	TP	SC
Δ^4, 3=CH$_2$	OH, CH$_3$	N-19	35	42	1	MT	O
Δ^4, 3=CH$_2$	OH, Et	N-20	4	5	1	MT	O
Δ^4, [2,3-d] isoxazole	—OAc	N-70	15	80	5	T	SC
Δ^4, [2,3-d] isoxazole	OH, CH$_3$	N-76	30	100	3	TP	SC
Δ^5, 3CO	OH, CH$_3$	N-39	Same as N-4				
Δ^5, 3CO	OH, Et	N-38	Same as N-5				
Δ^5, 3CO	OH, C≡CH	N-37	Same as N-7				
Δ^5, 3CO, 4,4,(CH$_3$)$_2$	OH, CH$_3$	N-42	0	0	—	TP	SC
Δ^5, 3βOH	OH, CH$_3$	N-26	50	700	14	MT	O
Δ^5, 3βOH	OH, Et	N-27	20	400	20	MT	O
Δ^5, 3CO, 4,4,(CH$_3$)$_2$	—OH	N-41	0	0	—	TP	SC
1α CH$_3$, 3CO	—OH	N-48	0	20	20	T	SC

291

Table V.2 *(continued)*

Compound	R	#	Andr	Anab	Q	St	Rte
1α CH₃, 3 CO —OAc		N-61	<1.5	20-30	15	TP	SC
1α OH, 3 CO —OAc		N-62	2-3	20-30	10	TP	SC
1α ,2α , —OAc		N-63	4.5	20-30	5	TP	SC
=CH₂, 3CO							
2,2 (CH₃)₂, ⁻OH							
3β OH ⋯CH₃		N-24	0	0	—	TP	SC
[2,3-*d*] ⁻OH							
isoxazole ⋯CH₃		N-72	10	10	1	TP	SC
2αCH₃, 3CO —OH		N-22	20	40	2	TP	SC
3CO —OH		N-89	—	<10	—	TP	SC
3CO ⁻OH ⋯CH₃		N-90	10	100	10	TP	SC
3CO ⁻OH ⋯Et		N-91	—	10	—	TP	SC
3CO, 5ξCH₃ —OH		N-25	<5	14	3	T	SC

Table V.3

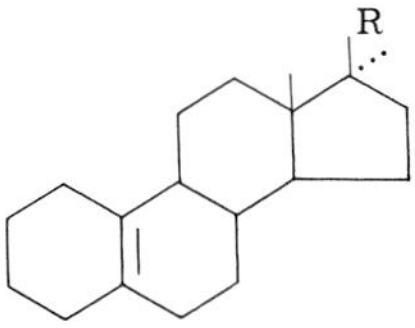

Compound	R	#	Andr	Anab	Q	St	Rte
—	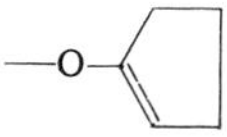 —O—⬠	N-71	40	120	3	MT	O
—	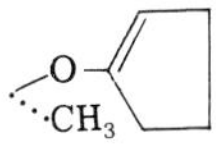 O—⬠ / ···CH₃	N-83	130	286	2	MT	O
3CO	—OH	N-92	—	5	—	TP	SC
3CO	OH / ···CH₃	N-40	25	150	6	T	SC
3CO	OH / ···Et	N-93	—	5	—	TP	SC
3CO	OH / ···C≡CH	N-43	<10	<10	—	T	SC

293

Table V.4

Compound	R_2	R_1	#	Andr	Anab	Q	St	Rte
Δ^2	CH_2OH	OH / CH$_3$	N-56	<1	1.3	—	T	SC
Δ^2	OH	OH / CH$_3$	N-49	8	14	<2	MT	O
Δ^2	CH_2OH	—OH	N-55	0	0	—	T, MT	SC,O
2,3β Epoxy	CN	—OAc	N-44	0	0	—	TP	SC
3CO,5F	OH	OH / C≡CH	N-69	—	—	1.7	T	SC
3CO,Δ^4	CN	=O	N-58	0	0	—	TP	SC
3CO,Δ^4	OH	—OH	N-68	—	—	1.5	T	SC
3CO Δ^4	Vinyl	—OH	N-100	80	50	0.6	T	SC
3CO Δ^4	Ethyl	—OH	N-101	4	4	—	T	SC
3β OH Δ^5	CHO	OH / CH$_3$	N-59	<10	<10	—	TP	SC
3β OH Δ^5	CH_2OH	OH / CH$_3$	N-60	<10	<10	—	TP	SC
3β OH Δ^5	CN	—OH	N-57	0	0	—	TP	SC

Table V.5

Compound	R	#	Andr	Anab	Q	St	Rte
Δ^4	⟋OH ⋯Et	N-47	120	160	1.3	EtNT	SC
$3CO, \Delta^4$	—OH	N-45	27	54	2	TP	SC
$3CO, \Delta^4$	⟋OH ⋯Et	N-46	17	350	20	TP	SC
$3CO, \Delta^4$	⟋OH ⋯C≡CH	N-51	7	70	10	TP	SC
$3CO, \Delta^4$, $7\alpha\,CH_3$	—OH	N-52	60	300	5	TP	SC
$3CO, \Delta^4$, $7\alpha\,CH_3$	⟋OH ⋯Et	N-53	70	48	0.7	TP	SC
$3CO, \Delta^4$, $7\alpha\,CH_3$	⟋OH ⋯C≡CH	N-54	26	27	1	TP	SC
$3CO, \Delta^4$, $\Delta^{9(10)}$	⟋OH ⋯Et	N-98	860	1450	2	EtNT	SC

Table V.6

Compound	R	#	Andr	Anab	Q	St	Rte
—	—OH	N-95	10	100	10	MT	SC
—	OH / CH$_3$	N-96	250	1000	4	MT	O
—	OH / CH$_3$	N-96	30	130	4	MT	SC
—	OH / Et	N-97	10	100	10	MT	SC
—	OH / Et	N-97	30	150	5	MT	O
Δ^{11}	OH / CH$_3$	N-99	8000	12000	1.5	MT	C
Δ^{11}, 7α CH$_3$	CH / CH$_3$	N-102	<10,000	<10,000	<1	MT	O
Δ^{11}	—OAc	N-103	500	500	1	NTAC	SC
Δ^{11}	—OCH$_2$ OCH$_3$	N-104	2000	2000	1	MT	O
Δ^{11}	—OCO C$_9$H$_{19}$	N-105	500	500	1	NTPP	SC
Δ^{11}	—OCO OCH$_2$Ph	N-106	500	500	1	NTPP	SC

BIBLIOGRAPHY

1. For review, see A. Albert, "Selective Toxicity," 3rd ed. Wiley, New York, 1965.
2. E. J. Ariens, B. M. Bloom, H. O. J. Collier, R. F. Furchgott, I. M. Goldman, D. Mackay, W. D. M. Paton, H. P. Rang, J. M. Van Rossum, and P. G. Wasser, *Advan. Drug Res.* **3** (1966).
3. H. J. Ringold, *in* "Mechanism of Action of Steroid Hormones" (C. A. Villee and L. L. Engel, eds.), p. 200. Pergamon Press, Oxford, 1961.
4. I. E. Bush, *Pharmacol. Rev.* **14**, 317 (1962).
5. H. L. Krüskemper, "Anabole Steroide," 2nd ed. Thieme, Stuttgart, 1965.
6. I. E. Bush, *Proc. 2nd Intern. Congr. Endocrinol., London, 1964.* Part 2, pp. 1324–1335. Excerpta Med. Found., Amsterdam, 1965.
7. I. E. Bush, *in* "Hormonal Steroids" (L. Martini, F. Fraschini, and M. Motta, eds.), Excerpta Med. Intern. Congr. Ser. No. 132, p. 60. Excerpta Med. Found., Amsterdam, 1967.
8. R. I. Dorfman and R. A. Shipley, "Androgens." Wiley, New York, 1956.
9. B. Camerino and G. Sala, *Progr. Drug Res.* **2**, 71 (1960).
10. R. I. Dorfman and F. Ungar, "Metabolism of Steroid Hormones." Academic Press, New York, 1965.
11. A. J. Manson, F. W. Stonner, H. C. Neumann, R. G. Christiansen, R. L. Clarke, J. H. Ackerman, D. F. Page, J. W. Dean, D. K. Phillips, G. O. Potts, A. Arnold, A. L. Beyler, and R. O. Clinton, *J. Med. Chem.* **6**, 1 (1963).
12. A. Ercoli, R. Gardi, and R. Vitali, *Chem. & Ind. (London)* p. 1284 (1962).
13. E. Caspi and D. Piatak, *Chem. & Ind. (London)* p. 1984 (1962).
14. J. P. Ruelas, J. Iriarte, F. Kincl, and C. Djerassi, *J. Org. Chem.* **23**, 1744 (1958).
15. S. Bernstein, S. M. Stollar, and M. Heller, *J. Org. Chem.* **22**, 472 (1957).
16. C. Djerassi, H. Bendas, and A. Segaloff, *J. Org. Chem.* **21**, 1056 (1956).
17. C. W. Marshall, J. W. Ralls, F. J. Saunders, and B. Riegel, *J. Biol. Chem.* **228**, 339 (1957).
18. J. A. Hogg, G. B. Spero, J. L. Thompson, B. J. Magerlein, W. P. Schneider, D. H. Peterson, O. K. Sebek, H. C. Murray, J. C. Babcock, R. L. Pederson, and J. A. Campbell, *Chem. & Ind. (London)* p. 1002 (1958).
19. C. D. Kochakian, *Proc. Soc. Exptl. Biol. Med.* **80**, 386 (1952).
20. R. I. Dorfman and D. Stevens, *Acta Endocrinol.* Suppl. 51, 867 (1960).
21. R. I. Dorfman, *Science* **131**, 1096 (1960).
22. A. Segaloff and R. B. Gabbard, *Endocrinology* **67**, 887 (1960).
23. H. J. Ringold, E. Batres, O. Halpern, and E. Necoechea, *J. Am. Chem. Soc.* **81**, 427 (1959).
24. B. J. Magerlein and J. A. Hogg, *J. Am. Chem. Soc.* **80**, 2220 (1958).

25. J. A. Campbell, J. C. Babcock, and J. A. Hogg, *J. Am. Chem. Soc.* **80**, 4717 (1958).
26. G. Sala, *Helv. Med. Acta* **27**, 519 (1960).
27. F. Sondheimer and Y. Mazur, *J. Am. Chem. Soc.* **79**, 2906 (1957).
28. J. A. Edwards and H. J. Ringold, *J. Am. Chem. Soc.* **81**, 5262 (1959).
29. R. E. Counsell and P. D. Klimstra, *J. Med. Pharm. Chem.* **5**, 477 (1962).
30. R. I. Dorfman and F. A. Kincl, *Endocrinology* **72**, 259 (1963).
31. V. A. Drill and B. Riegel, *Recent Progr. Hormone Res.* **14**, 29 (1958).
32. G. Sala and G. Baldratti, *Proc. Soc. Exptl. Biol. Med.* **95**, 22 (1957).
33. N. W. Atwater, *J. Am. Chem. Soc.* **82**, 2847 (1960).
34. H. J. Ringold, E. Batres, and G. Rosenkranz, *J. Org. Chem.* **22**, 99 (1957).
35. A. Segaloff, *Steroids* **1**, 299 (1963).
36. F. J. Saunders and V. A. Drill, *Proc. Soc. Exptl. Biol. Med.* **94**, 646 (1957).
37. S. C. Lyster, G. H. Lund, and R. O. Stafford, *Endocrinology* **58**, 781 (1956).
38. H. L. Krüskemper, *Arzneimittel-Forsch.* **16**, 608 (1966).
39. H. J. Ringold, E. Batres, O. Halpern, and E. Necoechea, *J. Am. Chem. Soc.* **81**, 427 (1959).
40. P. G. Holton and E. Necoechea, *J. Med. Pharm. Chem.* **5**, 1352 (1962).
41. G. O. Potts, A. L. Beyler, and D. F. Burnham, *Proc. Soc. Exptl. Biol. Med.* **103**, 383 (1960).
42. G. A. Overbeek, A. Delver, and J. de Visser, *Acta Endocrinol.* Suppl. 63, 7 (1961).
43. A. Bowers, A. D. Cross, J. A. Edwards, H. Carpio, M. C. Calzada, and E. Denot, *J. Med. Chem.* **6**, 156 (1963).
44. A. Arnold and G. O. Potts, *Acta Endocrinol.* **52**, 489 (1966).
45. K. Junkmann and G. Suchowsky, *Arzneimittel-Forsch.* **12**, 214 (1962).
46. R. E. Counsell, P. D. Klimstra, and F. B. Colton, *J. Org. Chem.* **27**, 248 (1962).
47. G. K. Suchowsky and K. Junkmann, *Klin. Wochschr.* **39**, 369 (1961).
48. K. Irmscher, H. G. Kraft, and K. Brückner, *J. Med. Chem.* **7**, 345 (1964).
49. A. Arnold, G. O. Potts, and A. L. Beyler, *Endocrinology* **72**, 408 (1963).
50. G. A. Overbeek, "Anabole Steroide." Springer, Berlin, 1966.
51. R. Pappo and C. J. Jung, *Tetrahedron Letters*, p. 365 (1962).
52. H. D. Lennon and F. J. Saunders, *Steroids* **4**, 689 (1964).
53. A. D. Cross, H. Carpio, and H. J. Ringold, *J. Med. Chem.* **6**, 198 (1963).
54. H. G. Lehmann, O. Engelfried, and R. Wiechert, *J. Med. Chem.* **8**, 383 (1965).
55. T. Feyel-Cabanes, *Compt. Rend. Soc. Biol.* **157**, 1428 (1963).
56. F. A. Kincl and R. I. Dorfman, *Acta Endocrinol.* **46**, 300 (1964).
57. A. D. Cross, J. A. Edwards, J. C. Orr, B. Berköz, L. Cervantes, M. C. Calzada, and A. Bowers, *J. Med. Chem.* **6**, 162 (1963).
58. J. C. Orr, O. Halpern, P. G. Holton, F. Alvarez, I. Delfin, A. de la Roz, A. M. Ruiz, and A. Bowers, *J. Med. Chem.* **6**, 166 (1963).
59. J. A. Edwards, P. G. Holton, J. C. Orr, L. C. Ibanez, E. Necoechea, A. de la Roz, E. Segovia, R. Urquiza, and A. Bowers, *J. Med. Chem.* **6**, 174 (1963).
60. F. A. Kincl, H. J. Ringold, and R. I. Dorfman, *Acta Endocrinol.* **36**, 83 (1961).
61. R. I. Dorfman and A. S. Dorfman, *Acta Endocrinol.* **42**, 245 (1963).
62. C. J. Meyer, C. Krähenbühl, and P. A. Desaulles, *Acta Endocrinol.* **43**, 27 (1963).
63. A. Arnold, G. O. Potts, and A. L. Beyler, *Acta Endocrinol.* **44**, 490 (1963).
64. P. A. Desaulles and C. Krähenbühl, *Acta Endocrinol.* **40**, 217 (1960).
65. L. E. Barnes, R. O. Stafford, M. E. Guild, L. C. Thole, and K. J. Olson, *Endocrinology* **55**, 77 (1954).
66. G. A. Overbeek and J. de Visser, *Acta Endocrinol.* **22**, 318 (1956).
67. F. J. Saunders and V. A. Drill, *Endocrinology* **58**, 567 (1956).

68. J. Iriarte, C. Djerassi, and H. J. Ringold, *J. Am. Chem. Soc.* **81,** 436 (1959); see H. J. Ringold, *Ann. N. Y. Acad. Sci.* **71,** 515 (1958).
69. D. L. Cook, R. A. Edgren, and F. J. Saunders, *Endocrinology* **62,** 798 (1958).
70. S. C. Lyster and G. W. Duncan, *Acta Endocrinol.* **43,** 399 (1963).
71. M. E. Wolff, W. Ho, and R. Kwok, *J. Med. Chem.* **7,** 577 (1964).
72. J. C. Stucki, G. W. Duncan, and S. C. Lyster, *Proc. Intern. Congr. Hormonal Steroids, Milan, 1962* Excerpta Med. Intern. Congr. Ser. No. 51, p. 65. Excerpta Med. Found., Amsterdam, 1962; *in* "Hormonal Steroids" (L. Martini and A. Pecile, eds.) Vol. 2, p. 119. Academic Press, New York, 1965.
73. J. A. Campbell, S. C. Lyster, G. W. Duncan, and J. C. Babcock, *Steroids* **1,** 317 (1963).
74. H. L. Krüskemper, K. D. Morgner, and G. Noell, *Arzneimittel-Forsch.* **17,** 449 (1967).
75. H. L. Krüskemper and G. Noell, *Steroids* **8,** 13 (1966).
76. H. G. Kraft and K. Brückner, *Arzneimittel-Forsch.* **14,** 328 (1964).
77. R. A. Edgren, D. L. Peterson, R. C. Jones, C. L. Nagra, H. Smith, and G. A. Hughes, *Recent Progr. Hormone Res.* **22,** 305 (1966).
78. A. Zaffaroni, *Acta Endocrinol.* Suppl. 50, 139 (1960).
79. J. W. Perrine, *Acta Endocrinol.* **37,** 376 (1961).
80. A. Segaloff and R. B. Gabbard, *Steroids* **1,** 77 (1963).
81. R. A. Edgren, *Acta Endocrinol.* **44,** Suppl., 87, pp. 3–21 (1963).
82. F. A. Kincl and R. I. Dorfman, *Steroids* **3,** 109 (1964).
83. S. D. Levine, *J. Med. Chem.* **8,** 537 (1965).
84. E. F. Nutting, P. D. Klimstra, and R. E. Counsell, *Acta Endocrinol.* **53,** 627 and 635 (1966).
85. P. D. Klimstra and R. E. Counsell, *J. Med. Chem.* **8,** 48 (1965).
86. P. D. Klimstra, R. Zigman, and R. E. Counsell, *J. Med. Chem.* **9,** 924 (1966).
87. P. D. Klimstra, E. F. Nutting, and R. E. Counsell, *J. Med. Chem.* **9,** 693 (1966).
88. G. C. Buzby, C. R. Walk, and H. Smith, *J. Med. Chem.* **9,** 782 (1966).
89. R. E. Counsell, G. W. Adelstein, P. D. Klimstra, and B. Smith, *J. Med. Chem.* **9,** 685 (1966).
90. R. E. Havranek, G. B. Hoey, and D. H. Baeder, *J. Med. Chem.* **9,** 326 (1966).
91. D. P. Strike, D. Herbst, and H. Smith, *J. Med. Chem.* **10,** 446 (1967).
92. P. E. Shaw, F. W. Gubitz, K. F. Jennings, G. O. Potts, A. L. Beyler, and R. L. Clarke, *J. Med. Chem.* **7,** 555 (1964).
93. J. H. Ackerman, G. O. Potts, A. L. Beyler, and R. O. Clinton, *J. Med. Chem.* **7,** 238 (1964).
94. J. A. Zderic, H. Carpio, A. Ruiz, D. C. Limon, F. Kincl, and H. J. Ringold, *J. Med. Chem.* **6,** 195 (1963).
95. S. Nakanishi, *J. Med. Chem.* **7,** 106 (1964)
96. S. Nakanishi, *J. Med. Chem.* **7,** 108 (1964).
97. R. E. Counsell and P. D. Klimstra, *J. Med. Chem.* **6,** 736 (1963).
98. J. A. Edwards, M. C. Calzada, and A. Bowers, *J. Med. Chem.* **6,** 178 (1963).
99. M. E. Wolff and T. Jen, *J. Med. Chem.* **6,** 726 (1963).
100. F. Neumann and R. Wiechert, *Arzneimittel-Forsch.* **15,** 1168 (1965).
101. M. Fox, A. S. Minot, and G. W. Liddle, *J. Clin. Endocrinol. Metab.* **22,** 921 (1962).
102. R. I. Dorfman, E. Caspi, and P. K. Grover, *Proc. Soc. Exptl. Biol. Med.* **110,** 750 (1962).
103. G. W. Duncan, S. C. Lyster, and J. A. Campbell, *Proc. Soc. Exptl. Biol. Med.* **116,** 800 (1964).
104. A. H. Goldkamp, *J. Med. Chem.* **5,** 1176 (1962).

105. C. Cavallero, *Acta Endocrinol.* **55,** 119 (1967).
106. C. Cavallero and P. Ofner, *Acta Endocrinol.* **55,** 131 (1967).
107. F. A. Kincl, *Methods Hormone Res.* **4,** 21 (1965).
108. P. de Ruggieri, R. Matscher, C. Gandolfi, D. Chiaramonti, C. Lupo, E. Pietra, and R. Cavalli, *Arch. Sci. Biol. (Bologna)* **47,** 1 (1963).
109. R. Hüttenrauch, *Arch. Pharm.* **298,** 20 (1965).
110. A. Schubert, A. Stachowiak, D. Onken, H. Specht, K. Barnikol-Oettler, E. Bode, K. Heller, W. Pohnert, S. Schwarz, and R. Zepter, *Pharmazie* **18,** 323 (1963).
111. H. J. Ringold, *J. Am. Chem. Soc.* **82,** 961 (1960).
112. M. W. Goldberg and R. Monnier, *Helv. Chim. Acta* **23,** 840 (1940).
113. M. W. Goldberg and H. Kirchensteiner, *Helv. Chim. Acta* **26,** 288 (1943).
114. T. I. Kornilova and G. L. Zhdanov, *Dokl. Akad. Nauk SSSR* **145,** 1163 (1962).
115. P. Ofner, H. H. Harvey, J. Sasse, P. L. Munson, and K. J. Ryan, *Endocrinology* **70,** 149 (1962).
116. J. M. H. Graves, G. A. Hughes, T. Y. Jen, and H. Smith, *J. Chem. Soc.* p. 5488 (1964).
117. L. Mamlok, A. Horeau, and J. Jacques, *Bull. Soc. Chim. France* p. 2359 (1965).
118. R. M. Scribner, *J. Org. Chem.* **30,** 3203 (1965).
119. P. Westerhof, J. Hartog, and S. J. Halkes, *Rec. Trav. Chim.* **84,** 863 (1965).
120. P. Westerhof and J. Hartog, *Rec. Trav. Chim.* **84,** 918 (1965).
121. M. Shimizu, G. Ohta, K. Veno, T. Takegoshi, Y. Oshima, A. Kasahara, T. Onodera, M. Mogi, and H. Tachizawa, *Chem. & Pharm. Bull. (Tokyo)* **13,** 895 (1965).
122. M. Shimizu, G. Ohta, K. Veno, and T. Takegoshi, *Chem. & Pharm. Bull. (Tokyo)* **12,** 77 (1964).
123. R. F. R. Church, A. S. Kende, and M. J. Weiss, *J. Am. Chem. Soc.* **87,** 2665 (1965).
124. A. L. Beyer, G. O. Potts, and A. Arnold, *Endocrinology* **68,** 987 (1961).
125. P. D. Klimstra and R. E. Counsell, *J. Med. Pharm. Chem.* **5,** 1216 (1962).
126. H. L. Saunders and M. C. Doggett, *Federation Proc.* **20,** 175e (1961).
127. G. Sala, *in* "Hormonal Steroids" (L. Martini and A. Pecile, eds.), Vol 1, p. 67. Academic Press, New York, 1964.
128. G. R. McKinney and H. G. Payne, *Proc. Soc. Exptl. Biol. Med.* **108,** 273 (1961).
129. R. Degenghi and R. Gaudry, *Can. J. Chem.* **40,** 818 (1962).
130. A. Arnold, G. O. Potts, A. L. Beyler, and R. O. Clinton. *Federation Proc.* **20,** 198 (1961).
131. P. Donini and R. Montezemolo, *Farmaco (Pavia), Ed. Sci.* **16,** 633 (1961).
132. J. H. Fried, A. N. Nutile, and G. E. Arth, *J. Am. Chem. Soc.* **82,** 5704 (1960).
133. R. E. Schaub and M. J. Weiss, *J. Org. Chem.* **26,** 3915 (1961).
134. C. H. Robinson, L. E. Finckenor, R. Tiberi, M. Eisler, R. Neri, A. Watnick, P. L. Perlman, P. Holroyd, W. Charney, and E. P. Oliveto, *J. Am. Chem. Soc.* **82,** 4611 (1960).
135. A. Bowers, P. G. Holton, E. Necoechea, and F. A. Kincl, *J. Chem. Soc.* p. 4057 (1961).
136. J. H. Fried, A. N. Nutile, G. E. Arth, and L. H. Sarett, *J. Org. Chem.* **27,** 682 (1962).
137. P. deRuggieri and C. Gandolfi, *in* "Hormonal Steroids" (L. Martini and A. Pecile, eds.), Vol. 2, p. 69. Academic Press, New York, 1965.
138. C. H. Robinson, O. Gnoj, and E. P. Oliveto, *J. Org. Chem.* **25,** 2247 (1960).
139. A. Ercoli, G. Bruni, G. Falconi, R. Gardi, and A. Meli, *Endocrinology* **67,** 521 (1960).
140. A. Ercoli and R. Gardi, *J. Am. Chem. Soc.* **82,** 746 (1960).
141. A. Meli, *Endocrinology* **72,** 715 (1963).
142. A. Meli, A. Lewis, and B. Steinetz, *Steroids* **1,** 287 (1963).

143. G. Falconi, *in* "Hormonal Steroids" (L. Martini and A. Pecile, eds.), Vol. 2, p. 143. Academic Press, New York, 1965.
144. R. Vitali, R. Gardi, G. Falconi, and A. Ercoli, *Steroids* **8**, 527 (1966).
145. A. D. Cross and I. T. Harrison, *Steroids* **6**, 397 (1965).
146. A. D. Cross, I. T. Harrison, P. Crabbé, F. A. Kincl, and R. I. Dorfman, *Steroids* **4**, 229 (1964).
147. R. T. Rapala, R. J. Kraay, and K. Gerzon, *J. Med. Chem.* **8**, 580 (1965).
148. J. de Visser and G. A. Overbeek, *Acta Endocrinol.* **35**, 405 (1960).
149. F. J. Saunders, *Proc. Soc. Exptl. Biol. Med.* **123**, 303 (1966).
150. G. Cooley, J. W. Ducker, B. Ellis, V. Petrow, and W. P. Scott, *J. Chem. Soc.* p. 4108 (1961).
151. J. C. Bloch, P. Crabbé, F. A. Kincl, G. Ourisson, J. Perez, and J. A. Zderic, *Bull. Soc. Chim. France* p. 559 (1961).
152. W. S. Johnson, M. Neeman, S. P. Birkeland, and N. A. Fedoruk, *J. Am. Chem. Soc.* **84**, 989 (1962).
153. M. Neeman, M. C. Caserio, J. D. Roberts, and W. S. Johnson, *Tetrahedron* **6**, 36 (1959).
154. M. Perelman, E. Farkas, E. J. Fornefeld, R. J. Kraay, and R. T. Rapala, *J. Am. Chem. Soc.* **82**, 2402 (1960).
155. A. Bowers, *in* "Hormonal Steroids" (L. Martini and A. Pecile, eds.), Vol. 2, p. 133. Academic Press, New York, 1965.
156. J. A. Campbell and J. C. Babcock, *in* "Hormonal Steroids" (L. Martini and A. Pecile, eds.), Vol. 2, p. 59. Academic Press, New York, 1965.
157. R. R. Burtner and R. E. Gentry, *J. Org. Chem.* **25**, 582 (1960).
158. L. J. Lerner, A. V. Bianchi, and M. Dzelzkalns, *Steroids* **6**, 215 (1965).
159. J. Mathieu, *Proc. Intern. Symp. Drug Res., 1967* p. 134. Chem. Inst. Can., Montreal, Canada, 1967.
160. W. Nagata, *Proc. Intern. Symp. Drug. Res., 1967* p. 188. Chem. Inst. Can., Montreal, Canada, 1967.
161. J. A. Zderic, E. Batres, D. C. Limon, H. Carpio, J. Lisci, G. Monroy, E. Necoechea, and H. J. Ringold, *J. Am. Chem. Soc.* **82**, 3404 (1960).
162. W. S. Johnson, H. Lemaire, and R. Pappo, *J. Am. Chem. Soc.* **75**, 4866 (1953).
163. D. Magrath, V. Petrow, and R. Royer, *J. Chem. Soc.* p. 845 (1951).
164. K. Miescher and W. Klarer, *Helv. Chim. Acta* **22**, 962 (1939).
165. H. Heusser, N. Wahba, and F. Winternitz, *Helv. Chim. Acta* **37**, 1052 (1954).
166. C. D. Kochakian and J. R. Murlin, *J. Nutr.* **10**, 437 (1935).
167. C. D. West, F. H. Tyler, H. Brown, and L. T. Samuels, *J. Clin. Endocrinol.* **11**, 897 (1951).
168. G. M. Tomkins, *J. Biol. Chem.* **225**, 13 (1957).
169. E. E. Baulieu and P. Mauvais-Jarvis, *J. Biol. Chem.* **239**, 1569 and 1578 (1964).
170. S. G. Korenman and H. Wilson, *Steroids* **8**, 729 (1966).
171. C. D. Kochakian. *Proc. 4th Intern. Congr. Biochem., Vienna, 1958* Vol. 4, p. 196. Pergamon Press, Oxford, 1959.
172. C. J. Migeon, O. L. Lescure, W. H. Zinkman, and J. B. Sidbury, *J. Clin. Invest.* **41**, 2025 (1962).
173. H. H. Wotiz and H. M. Lemon, *J. Biol. Chem.* **206**, 525 (1954).
174. E. Gurpide, P. MacDonald, A. Chapdelaine, R. L. Vande Wiele, and S. Lieberman, *J. Clin. Endocrinol. Metab.* **25**, 1537 (1965).
175. P. C. MacDonald, A. Chapdelaine, O. Gonzales, E. Gurpide, R. L. Vande Wiele, and S. Lieberman, *J. Clin. Endocrinol. Metab.* **25**, 1557 (1965).

176. A. Chapdelaine, P. C. MacDonald, O. Gonzales, E. Gurpide, R. L. Vande Wiele, and S. Lieberman, *J. Clin. Endocrinol. Metab.* **25,** 1569 (1965).

177. L. L. Engel, J. Alexander, and M. Wheeler, *J. Biol. Chem.* **231,** 159 (1958).

178. G. A. Overbeek, J. vander Vies, and J. de Visser, *in* "Protein Metabolism" (F. Gross, ed.), pp. 189–193. Springer, Berlin, 1962.

179. For a summary, see A. Ercoli, R. Gardi, and G. Bruni, *Res. Progr. Org-Biol. Med. Chem.* **1,** 155 (1964).

180. L. H. Sarett, *Ann. N. Y. Acad. Sci.* **82,** 802 (1959).

181. W. D. M. Paton, *Proc. Roy. Soc.* **B154,** 21 (1961).

182. R. I. Dorfman, *Methods Hormone Res.* **2,** 275 (1962).

183. R. I. Dorfman, *in* "Hormone Assay" (C. W. Emmens, ed.), p. 291. Academic Press, New York, 1950.

184. A. Arnold, A. L. Beyler, and G. O. Potts, *Proc. Soc. Exptl. Biol. Med.* **102,** 184 (1959).

185. E. Eisenberg and G. S. Gordan, *J. Pharmacol. Exptl. Therap.* **99,** 38 (1950).

186. K. J. Hayes, *Acta Endocrinol.* **48,** 332 (1965).

187. L. G. Hershberger, E. G. Shipley, and R. K. Meyer, *Proc. Soc. Exptl. Biol. Med.* **83,** 175 (1953).

188. C. Huggins and E. V. Jensen, *J. Exptl. Med.* **100,** 241 (1954).

189. C. D. Kochakian, *Am. J. Physiol.* **158,** 51 (1949).

190. B. G. McFarland, *in* "Steroid Reactions" (C. Djerassi, ed.), p. 427. Holden-Day, San Francisco, California, 1963.

191. R. O. Stafford, B. J. Bowman, and K. J. Olson, *Proc. Soc. Exptl. Biol. Med.* **86,** 322 (1954).

192. J. C. Stucki, A. D. Forbes, J. I. Notham, and J. J. Clark, *Endocrinology* **66,** 585 (1960).

193. A. J. Birch, *J. Chem. Soc.* p. 367 (1950).

194. M. W. Goldberg and E. Wydler, *Helv. Chim. Acta* **26,** 1142 (1943).

195. A. L. Wilds, C. H. Shunk, and C. H. Hoffmann, *J. Am. Chem. Soc.* **71,** 3266 (1949).

196. C. Djerassi, R. Riniker, and B. Riniker, *J. Am. Chem. Soc.* **78,** 6362 (1956).

197. P. Westerhof and E. H. Reerink, *Rec. Trav. Chim.* **79,** 771 and 795 (1960).

198. E. H. Reerink, H. F. L. Schöler, P. Westerhof, A. Querido, A. A. H. Kassenaar, E. Diczfalusy, and K. C. Tillinger, *Nature* **186,** 168 (1960).

199. R. van Moorselaar, S. J. Halkes, and E. Havinga, *Rec. Trav. Chim.* **84,** 841 (1965).

200. P. Westerhof and A. Smit, *Rec. Trav. Chim.* **80,** 1048 (1961).

201. H. F. L. Schöler and A. M. de Wachter, *Acta Endocrinol.* **38,** 128 (1961).

202. J. P. L. Bots, *Rec. Trav. Chim.* **77,** 1010 (1958).

203. R. O. Clinton, A. J. Manson, F. W. Stonner, A. L. Beyler, G. O. Potts, and A. Arnold, *J. Am. Chem. Soc.* **81,** 1513 (1959).

204. R. O. Clinton, A. J. Manson, F. W. Stonner, H. C. Neumann, R. G. Christiansen, R. L. Clarke, J. H. Ackerman, D. F. Page, J. W. Dean, W. B. Dickinson, and C. Carabateas, *J. Am. Chem. Soc.* **83,** 1478 (1961).

205. R. O. Clinton, A. J. Manson, F. W. Stonner, R. G. Christiansen, A. L. Beyler, G. O. Potts, and A. Arnold, *J. Org. Chem.* **26,** 279 (1961).

206. R. G. Christiansen, U.S. Patent 3,287,355 (Sterling Drug Inc.) (1966).

207. F. B. Colton and I. Laos, U.S. Patent 2,999,092 (G. D. Searle & Co.) (1961).

208. P. de Ruggieri, C. Gandolfi, and D. Chiaramonti, *Gazz. Chim. Ital.* **92,** 768 (1962).

209. Ciba, U.S. Patent 3,033,860 (1962).

210. L. L. Smith, D. M. Teller, and T. Foell, *J. Med. Chem.* **6,** 330 (1963).

211. J. A. Vida and M. Gut, *J. Med. Chem.* **6**, 792 (1963).
212. J. A. Vida and M. Gut, *Steroids* **2**, 499 (1963).
213. E. Caspi and D. M. Piatak, *Experientia* **19**, 465 (1963).
214. C. H. Robinson, L. E. Finckenor, R. Tiberi, and E. P. Oliveto, *Steroids* **3**, 639 (1964).
215. R. B. Woodward, *in* "Mechanism of Action of Steroid Hormones" (C. A. Villee and L. L. Engel, eds.), p. 197. Pergamon Press, Oxford, 1961.
216. W. Moffitt, R. B. Woodward, A. Moscowitz, W. Klyne, and C. Djerassi, *J. Am. Chem. Soc.* **83**, 4013 (1961).
217. C. C. Costain and B. P. Stoicheff, *J. Chem. Phys.* **30**, 777 (1959).
218. L. C. Krisher and E. B. Wilson, *J. Chem. Phys.* **31**, 882 (1959).
219. L. Pauling, "The Nature of the Chemical Bond," Chapter 4/8. Cornell Univ. Press, Ithaca, New York, 1960.
220. M. Hanack, "Conformation Theory," p. 146. Academic Press, New York, 1965.
221. G. M. Tomkins, *Recent Progr. Hormone Res.* **12**, 125 (1956).
222. L. Pauling, "The Nature of the Chemical Bond," Chapter 7/12. Cornell Univ. Press, Ithaca, New York, 1960.
223. I. E. Bush and V. B. Mahesh, *Biochem. J.* **71**, 718 (1959).
224. G. W. Liddle, J. E. Richard, and G. M. Tomkins, *Metab. Clin. Exptl.* **5**, 384 (1956).
225. B. Bagett, L. L. Engel, K. Savard, and R. I. Dorfman, *Federation Proc.* **14**, 175 (1955).
226. A. S. Meyer, *Biochim. Biophys. Acta* **17**, 440 (1955).
227. K. J. Ryan, *Acta Endocrinol.* **35**, Suppl. 51, 697 (1960).
228. N. Kuji, *Acta Endocrinol.* **37**, 71 (1961).
229. E. M. Glenn, S. L. Richardson, and B. J. Bowman, *Metab. Clin. Exptl.* **8**, 265 (1959).
230. A. Wettstein, *Helv. Chim. Acta* **23**, 388 (1940).
231. C. P. Balant and M. Ehrenstein, *J. Org. Chem.* **17**, 1587 (1932).
232. H. J. Ringold, E. Batres, A. Bowers, J. Edwards, and J. Zderic, *J. Am. Chem. Soc.* **81**, 3485 (1959).
233. H. Breuer, *Acta Endocrinol.* **40**, 111 (1962).
234. G. Dörner, F. Stahl, and R. Zabel, *Endokrinologie* **45**, 121 (1963).
235. A. A. Sandberg, W. R. Slaunwhite, and H. N. Antoniades, *Recent Progr. Hormone Res.* **13**, 209 (1957).
236. B. H. Levedahl and L. T. Samuels, *J. Biol. Chem.* **186**, 857 (1950).
237. S. Pearson and T. H. McGavack, *J. Clin. Endocrinol. Metab.* **14**, 472 (1954).
238. R. G. A. van Wayjen, *Helv. Med. Acta* **27**, 523 (1960).
239. G. Sala and E. Castegnaro, *Folia Endocrinol. (Pisa)* **11**, 348 (1958).
240. H. J. Ringold, S. Ramachandran, and E. Forchielli, *J. Biol. Chem.* **237**, PC260 (1962).
241. R. I. Dorfman, *Recent Progr. Hormone Res.* **22**, 272 (1966) (discussion).
242. G. A. Overbeek, "Anabole Steroide," p. 42. Springer, Berlin, 1966.
243. G. Sala, G. Baldratti, R. Ronchi, V. Clini, and C. Bertazzoli, *Sperimentale* **106**, 490 (1956).
244. G. Sala and G. Baldratti, *Endocrinology* **72**, 494 (1963).
245. J. W. Partridge, L. Boling, L. DeWind, S. Margen, and L. W. Kinsell, *J. Clin. Endocrinol. Metab.* **13**, 189 (1953).
246. G. A. Overbeek and J. de Visser, *Acta Endocrinol.* **24**, 209 (1957).
247. I. Weis, *Wien. Med. Wochschr.* **112**, 230 (1962).
248. G. A. Overbeek and J. de Visser, *Acta Endocrinol.* **38**, 285 (1961).
249. E. Diczfalusy, *Acta Endocrinol.* **35**, 59 (1960).
250. F. J. Saunders and V. A. Drill, *Metab., Clin. Exptl.* 315 (1958).

251. H. Spencer, E. Berger, M. L. Charles, E. D. Gottesman, and D. Laszlo, *J. Clin. Endocrinol. Metab.* **17,** 975 (1957).
252. F. J. Saunders, F. B. Colton, and V. A. Drill, *Proc. Soc. Exptl. Biol. Med.* **94,** 717 (1957).
253. G. Pincus, M. C. Chang, M. X. Zarrow, E. S. E. Hafez, and A. Merrill, *Endocrinology* **59,** 695 (1956).
254. R. E. Edgren, *Acta Endocrinol.* **25,** 365 (1957).
255. J. N. Goldman, J. A. Epstein, and H. S. Kupperman, *Endocrinology* **61,** 166 (1957).
256. F. Plum and M. F. Dunning, *J. Clin. Endocrinol. Metab.* **18,** 860 (1958).
257. R. C. Kory, M. H. Bradley, R. N. Watson, R. Callahan, and B. J. Peters, *Am. J. Med.* **26,** 243 (1959).
258. A. Ercoli, *in* "Hormonal Steroids" (L. Martini and A. Pecile, eds.), Vol. 1, p. 11. Academic Press, New York, 1964.
259. G. Sala, G. Baldratti, and R. Ronchi, *Folia Endocrinol. (Pisa)* **10,** 729 (1957).
260. J. Bekaert, G. Deltour, R. Demol, and J. De Backer, *Arch. Studio Fisiopatol. Clin. Ricambio* **22,** 408 (1958).
261. A. Amerio, O. Micelli, and B. Paolicelli, *Arch. Studio Fisiopatol. Clin. Ricambio* **22,** 446 (1958).
262. L. Adezati, G. Lotti, and A. Polleri, *Arch. Studio Fisiopatol. Clin. Ricambio* **22,** 392 (1958).
263. G. Dörner and A. Schubert, *Proc. Intern. Congr. Hormonal Steroids, Milan, 1962,* Excerpta Med. Intern. Congr. Ser. No. 51, p. 210. Excerpta Med. Found., Amsterdam, 1962.
264. P. A. Desaulles, C. Krähenbühl, W. Schuler, and H. J. Bein, *Schweiz. Med. Wochschr.* **89,** 1313 (1959).
265. H. Breuer and U. Schikowski, *Acta Endocrinol.* **42,** 135 (1963).
266. G. K. Suchowsky and K. Junkmann, *Acta Endocrinol.* **39,** 68 (1962).
267. G. S. Gordon, *J. Am. Med. Assoc.* **162,** 600 (1956).
268. A. Segaloff, C. Y. Bowers, E. L. Rongone, P. J. Murison, and J. V. Schlosser, *Cancer* **11,** 1187 (1958).
269. J. A. Campbell and J. C. Babcock, *J. Am. Chem. Soc.* **81,** 4069 (1959).
270. H. L. Krüskemper and H. Breuer, *Proc. Intern. Congr. Hormonal Steroids, Milan, 1962.* Excerpta Med. Intern. Congr. Ser. No. 51, p. 209. Excerpta Med. Found., Amsterdam, 1962.
271. H. L. Krüskemper, *Klin. Wochschr.* **44,** 1127 (1966).
272. W. Jochle and H. Langecker, *Arzneimittel-Forsch.* **12,** 218 (1962).
273. K. H. Kimbel, K. H. Kolb, and P. E. Schulze, *Arzneimittel-Forsch.* **12,** 223 (1962).
274. K. H. Kolb, K. H. Kimbel, and P. E. Schulze, *Arzneimittel-Forsch.* **12,** 228 (1962).
275. H. Langecker, *Arzneimittel-Forsch.* **12,** 231 (1962).
276. O. Weller, *Arzneimittel-Forsch.* **12,** 234 (1962).
277. R. Wiechert and E. Caspar, *Chem. Ber.* **93,** 1710 (1960).
278. A. Popper and R. Wiechert, *Arzneimittel-Forsch.* **12,** 213 (1962).
279. M. S. deWinter, C. M. Siegman, and S. A. Szpilfogel, *Chem. & Ind. (London)* p. 905 (1959).
280. G. A. Overbeek, A. Delver, and J. de Visser, *Acta Endocrinol.* **40,** 133 (1962).
281. R. G. A. van Wayjen and G. Buyze, *Acta Endocrinol.* Suppl. 63, 18 (1962).
282. R. P. Howard and R. H. Furman, *J. Clin. Endocrinol. Metab.* **22,** 43 (1962).
283. E. Marchetti and P. Donini, *Gazz. Chim. Ital.* **86,** 1133 (1961).
284. E. Bertolotti and G. Lojodice, *Minerva Med.* **52,** 3433 (1961).

285. F. M. Antonini and G. Verdi, *Minerva Med.* **52**, 3437 (1961).

286. G. O. Potts and A. L. Beyler, *Proc. Intern. Congr. Hormonal Steroids, Milan, 1962.* Excerpta Med. Intern. Congr. Ser. No. 51, p. 211. Excerpta Med. Found., Amsterdam, 1962.

287. A. Arnold, G. O. Potts, and A. L. Beyler, *Federation Proc.* **21**, 212 (1962).

288. G. A. Overbeek, "Anabole Steroide," p. 51. Springer, Berlin, 1966.

289. R. A. Edgren and H. Smith, *Proc. Intern. Congr. Hormonal Steroids, Milan, 1962.* Excerpta Med. Intern. Congr. Ser. No. 51, p. 68. Excerpta Med. Found., Amsterdam, 1962.

290. R. M. Tomarelli and F. W. Bernhart, *Steroids* **4**, 451 (1964).

291. R. B. Greenblatt, E. C. Jungck, and G. C. King, *Am. J. Med. Sci.* **248**, 317 (1964).

292. H. Smith, G. A. Hughes, G. H. Douglas, D. Hartley, B. J. McLoughlin, J. B. Siddall, G. R. Wendt, G. C. Buzby, D. R. Herbst, K. W. Ledig, J. R. McMenamin, T. W. Pattison, J. Suida, J. Tokolics, R. A. Edgren, A. B. A. Jansen, B. Gadsby, D. H. R. Watson, and P. C. Philips, *Experientia* **19**, 394 (1963).

293. L. Velluz, G. Nominé, R. Bucourt, and J. Mathieu, *Compt. Rend.* **257**, 569 (1963).

294. M. Herrmann and H. G. Goslar, *Experientia* **19**, 76 (1963).

295. J. H. Peters, A. H. Randall, J. Mendeloff, R. Peace, J. C. Coberly, and M. B. Hurley, *J. Clin. Endocrinol. Metab.* **18**, 114 (1958).

296. H. A. Kaupp and F. W. Preston, *J. Am. Med. Assoc.* **180**, 411 (1962).

297. E. F. Gilbert, A. Q. DaSilva, and D. M. Queen, *J. Am. Med. Assoc.* **185**, 538 (1963).

298. W. J. Hogarth, *J. Can. Med. Assoc.* **88**, 368 (1963).

299. G. H. Marquardt, C. I. Fisher, P. Levy, and R. M. Dowben, *J. Am. Med. Assoc.* **175**, 851 (1961).

300. H. L. Krüskemper, *Klin. Wochschr.* **44**, 1127 (1966).

301. R. I. Dorfman, *Proc. 4th Intern. Congr. Biochemi., Vienna, 1958,* Vol. 4, p. 175. Pergamon Press, Oxford, 1959.

302. H. L. Krüskemper and G. Noell, *Acta Endocrinol.* **54**, 73 (1967).

303. K. D. Roberts, R. L. Vande Wiele, and S. Lieberman, *J. Biol. Chem.* **236**, 2213 (1961).

304. R. L. Vande Wiele, P. C. MacDonald, E. Gurpide, and S. Lieberman, *Recent Progr. Hormone Res.* **19**, 275 (1963).

305. M. B. Lipsett, S. G. Korenman, H. Wilson, and C. W. Bardin, *in* "Steroid Dynamics" (G. P. Pincus, J. F. Tait, and T. Nakao, eds.), p. 117. Academic Press, New York, 1966.

306. E. Gurpide, P. C. MacDonald, R. L. Vande Wiele, and S. Lieberman, *J. Clin. Endocrinol. Metab.* **23**, 346 (1963).

307. E. E. Baulieu, *Compt. Rend.* **251**, 1421 (1960).

308. E. E. Baulieu, *J. Clin. Endocrinol. Metab.* **22**, 501 (1962).

308a. E. E. Baulieu, *Recent Progr. Hormone Res.* **19**, 306 (1962) (discussion).

309. R. Hüttenrauch, *Pharmazie* **22**, 179 (1967).

310. A. A. Sandberg and W. R. Slaunwhite, *J. Clin. Invest.* **35**, 1331 (1956).

311. E. E. Baulieu, *in* "Research on Steroids" (C. Cassano, ed.), Vol. 1, p. 1. Il Pensiero Sci. Publ. Co., Rome, 1964.

312. E. E. Baulieu, C. Corpechot, F. Dray, R. Emiliozzi, M. C. Lebeau, P. Mauvais-Jarvis, and P. Robel, *Recent Progr. Hormone Res.* **21**, 411 (1965).

313. J. H. Richards and J. B. Hendrickson, "The Biosynthesis of Steroids, Terpenes and Acetogenins," Chapter 13. Benjamin, New York, 1964.

314. M. Hayano, M. Gut, R. I. Dorfman, A. Schubert, and R. Siebert, *Biochim. Biophys. Acta* **32**, 269 (1959).

315. M. Hayano, M. Gut, R. I. Dorfman, O. K. Sebek, and D. H. Peterson, *J. Am. Chem. Soc.* **80**, 2336 (1958).

316. S. Bergstrom, S. Lindstredt, B. Samuelson, E. J. Corey, and G. A. Gregoriou, *J. Am. Chem. Soc.* **80**, 2337 (1958).

317. E. J. Corey, G. A. Gregoriou, and D. H. Peterson, *J. Am. Chem. Soc.* **80**, 2338 (1958).

318. J. F. Tait, *J. Clin. Endocrinol. Metab.* **23**, 1285 (1963).

319. T. F. Gallagher, H. L. Bradlow, D. K. Fukushima, C. T. Beer, T. H. Kritchevsky, M. Stokem, M. L. Eidinoff, L. Hellman, and K. Dobriner, *Recent Progr. Hormone Res.* **9**, 411 (1954).

320. W. H. Pearlman, M. R. J. Pearlman, and A. E. Rakoff, *J. Biol. Chem.* **209**, 803 (1954).

321. R. E. Peterson, *Recent Progr. Hormone Res.* **15**, 231 (1959).

322. J. F. Tait and R. Horton, *Steroids* **4**, 365 (1964).

323. M. B. Lipsett, H. Wilson, M. A. Kirschner, S. G. Korenman, L. M. Fishman, G. A. Sarfaty, and C. W. Bardin, *Recent Progr. Hormone Res.* **22**, 245 (1966).

324. J. F. Tait and R. Horton, *in* "Steroid Dynamics" (G. Pincus, J. F. Tait, and T. Nakao, eds.), p. 393. Academic Press, New York, 1966.

325. S. Lieberman and E. Gurpide, *in* "Steroid Dynamics" (G. Pincus, J. F. Tait, and T. Nakao, eds.), p. 531. Academic Press, New York, 1966.

326. R. I. Dorfman, *Metab., Clin. Exptl.* **10**, 902 (1961).

327. K. J. Isselbacher, *Recent Progr. Hormone Res.* **12**, 134 (1956).

328. P. L. Munson, T. F. Gallagher, and F. C. Koch, *J. Biol. Chem.* **152**, 67 (1944).

329. H. I. Hadler, *Experientia* **11**, 175 (1955).

330. L. F. Fieser and M. Fieser, "Steroids," Chapter 7/6, p. 271. Reinhold, New York, 1959.

331. D. H. R. Barton, *Experientia* **6**, 316 (1950).

332. E. L. Eliel, N. L. Allinger, S. J. Angyal, and G. A. Morrison, "Conformational Analysis." Wiley (Interscience), New York, 1965.

333. M. J. T. Robinson, *Tetrahedron* **1**, 49 (1957).

334. R. P. Linstead, W. E. Doering, S. B. David, P. Levine, and R. R. Whetstone, *J. Am. Chem. Soc.* **64**, 1985 (1942).

335. M. C. Dart and H. B. Henbest, *J. Chem. Soc.* p. 3563 (1960).

336. C. W. Shoppee and T. Reichstein, *Helv. Chim. Acta* **24**, 351 (1941).

337. C. W. Shoppee, "Chemistry of the Steroids," p. 76. Academic Press, New York, 1958.

338. J. Pataki, G. Rosenkranz, and C. Djerassi, *J. Biol. Chem.* **195**, 791 (1952).

339. O. Mancera, H. J. Ringold, C. Djerassi, G. Rosenkranz, and F. Sondheimer, *J. Am. Chem. Soc.* **75**, 1286 (1953).

340. D. H. R. Barton and C. H. Robinson, *J. Chem. Soc.* p. 3045 (1954).

341. G. Stork and S. D. Darling, *J. Am. Chem. Soc.* **86**, 1761 (1964).

342. R. I. Dorfman, *Recent Progr. Hormone Res.* **9**, 5 (1954).

343. H. L. Bradlow and T. F. Gallagher, *J. Biol. Chem.* **229**, 505 (1957).

344. H. L. Bradlow and T. F. Gallagher, *J. Clin. Endocrinol. Metab.* **19**, 1575 (1959).

345. K. Savard, S. Burstein, H. Rosenkrantz, and R. I. Dorfman, *J. Biol. Chem.* **202**, 717 (1953).

346. D. H. R. Barton, *J. Chem. Soc.,* p. 1027 (1953).

347. D. H. R. Barton and R. C. Cookson, *Quart. Rev. (London)* **10**, 44 (1956).

348. E. L. Eliel, "Stereochemistry of Carbon Compounds," Chapter 8/7. McGraw-Hill, New York, 1962.
349. R. J. Wicker, *J. Chem. Soc.* p. 2165 (1956).
350. R. J. Wicker, *J. Chem. Soc.* p. 3299 (1957).
351. W. G. Dauben, G. J. Fonken, and D. S. Noyce, *J. Am. Chem. Soc.* **78,** 2579 (1956).
352. D. M. S. Wheeler and J. W. Huffman, *Experientia* **16,** 516 (1960).
353. H. Haubenstock and E. L. Eliel, *J. Am. Chem. Soc.* **84,** 2363 and 2368 (1962).
354. C. D. Ritchie and A. L. Pratt, *J. Am. Chem. Soc.* **86,** 1571 (1964).
355. H. C. Brown, "Hydroboration," Chapter 17. Benjamin, New York, 1962.
356. B. L. Rubin, *J. Biol. Chem.* **227,** 917 (1957).
357. H. C. Brown and J. Muzzio, *J. Am. Chem. Soc.* **88,** 2811 (1966).
358. B. M. Bloom and G. M. Shull, *J. Am. Chem. Soc.* **77,** 5767, 1955.
359. B. M. Bloom, quoted by M. Hayano, *in* "Oxygenases" (O. Hayaishi, ed.), p. 222. Academic Press, New York, 1962. The description is given in a modified form.
360. C. A. Coulson and E. T. Stewart, *in* "The Chemistry of Alkenes" (S. Patai, ed.), pp. 1–147. Wiley (Interscience), New York, 1964.
361. E. Cartmell and G. W. Fowles, "Valency and Molecular Structure," 2nd ed., Chapter 9. Academic Press, New York, 1961.
362. A. Rosowsky, *in* "Heterocyclic Compounds with Three- and Four-Membered Rings" (A. Weissberger, ed.), Part 1, Chapter 1. Wiley (Interscience), New York, 1964.
363. N. N. Schwartz and J. H. Blumbergs, *J. Org. Chem.* **29,** 1976 (1964).
364. M. Hayano, M. C. Lindberg, R. I. Dorfman, J. E. H. Hancock, and W. von E. Doering, *Arch. Biochem. Biophys.* **59,** 529 (1955).
365. M. Hayano, A. Saito, D. Stone, and R. I. Dorfman, *Biochim. Biophys. Acta* **21,** 380 (1956).
366. H. S. Mason, *Advan. Enzymol.* **19,** 79 (1957).
367. R. Breslow, "Organic Reaction Mechanisms," Chapter 6. Benjamin, New York, 1965.
368. J. Hine, "Physical Organic Chemistry," Chapter 10. McGraw-Hill, New York, 1956.
369. H. J. Ringold, quoted by M. Hayano, *in* "Oxygenases" (O. Hayaishi, ed.), p. 225. Academic Press, New York, 1962.
370. M. Hayano, *in* "Oxygenases" (O. Hayasishi, ed.), Chapter 5, p. 227. Academic Press, New York, 1962.
371. M. Hayano, J. E. Longchampt, W. Kelly, C. Gual, and R. I. Dorfman, *Acta Endocrinol.* **35,** Suppl. 51, 699 (1960).
372. J. E. Longchampt, M. Hayano, M. Ehrenstein, and R. I. Dorfman, *Endocrinology* **67,** 843 (1960).
373. H. Kwart and D. M. Hoffman, *J. Org. Chem.* **31,** 419 (1966).
374. D. J. Cram, J. L. Mateos, F. Hauck, A. Langemann, K. R. Kopecky, W. D. Nielsen, and J. Allinger, *J. Am. Chem. Soc.* **81,** 5774 (1959).
375. D. J. Cram, C. A. Kingsbury, and B. Rickborn, *J. Am. Chem. Soc.* **83,** 3688 (1961).
376. D. J. Cram and G. S. Hammond, "Organic Chemistry," 2nd ed., Chapters 14 and 20. McGraw-Hill, New York, 1964.
377. C. Gual, T. Morato, M. Hayano, M. Gut, and R. I. Dorfman, *Endocrinology* **71,** 920 (1962).
378. J. Fried, *Proc. Intern. Symp. Drug Res., 1967,* p. 206. Chemical Inst. Can., Montreal, Canada, 1967.
379. H. Gibian, *Proc. Intern. Symp. Drug Res., 1967,* p. 159. Chemical Inst. Can., Montreal, Canada, 1967.

380. L. Kohout, J. Fajkos, and F. Sorm, *in* "Hormonal Steroids" (L. Martini, F. Fraschini, and M. Motta, eds.), Excerpta Med. Intern. Congr. Ser. No. 132, p. 238. Excerpta Med. Found., Amsterdam, 1967.

381. J. van der Vies, *Acta Endocrinol.* **49,** 271 (1965).

382. G. A. Overbeek, I. L. Bonta, C. J. de Vos, J. de Visser, and J. van der Vies, *in* "Hormonal Steroids" (L. Martini, F. Fraschini, and M. Motta, eds.), Excerpta Med. Intern. Congr. Ser. No. 132, p. 68. Excerpta Med. Found., Amsterdam, 1967.

383. R. L. Vande Wiele and S. Lieberman, *in* "Biological Activities of Steroids in Relation to Cancer" (G. Pincus and E. P. Vollmer, eds.), p. 93. Academic Press, New York, 1960.

384. E. Gurpide, J. Mann, R. L. Vande Wiele, and S. Lieberman, *Acta Endocrinol.* **39,** 213 (1962).

385. E. E. Baulieu, *in* "Hormonal Steroids" (L. Martini, F. Fraschini, and M. Motta, eds.), Excerpta Med. Intern. Congr. Ser. No. 132, p. 37. Excerpta Med. Found., Amsterdam, 1967.

386. K. R. Laumas, J. F. Tait, and S. A. S. Tait, *Acta Endocrinol.* **36,** 265 (1961).

387. K. R. Laumas, J. F. Tait, and S. A. S. Tait, *Acta Endocrinol.* **38,** 469 (1961).

388. S. G. Korenman and M. B. Lipsett, *J. Clin. Invest.* **43,** 2125 (1964).

389. J. F. Tait and S. Burstein, *Hormones* **5,** 551 (1964).

390. S. Lieberman, *in* "Hormonal Steroids" (L. Martini, F. Fraschini, and M. Motta, eds.), Excerpta Med. Intern. Congr. Ser. No. 132, p. 22. Excepta Med. Found., Amsterdam, 1967.

391. L. L. Engel, *in* "Hormonal Steroids" (L. Martini, F. Fraschini, and M. Motta, eds.), Excerpta Med. Intern. Congr. Ser. No. 132, p. 52. Excerpta Med. Found., Amsterdam, 1967.

392. J. W. Dean, G. O. Potts, and R. G. Christiansen, *J. Med. Chem.* **10,** 795 (1967).

393. S. R. Landor and J. P. Regan, *J. Chem. Soc., C, Org.,* p. 1159 (1967).

394. A. Kasahara, T. Onodera, M. Mogi, Y. Oshima, and M. Shimizu, *Chem. & Pharm. Bull. (Tokyo)* **13,** 1460 (1965).

395. K. Veno and G. Ohta, *Chem. & Pharm. Bull. (Tokyo)* **15,** 518 (1967).

396. E. Shapiro, L. Weber, E. P. Oliveto, H. L. Herzog, S. Neri, S. Tolksdorf, M. Tanabe, and D. F. Crowe, *Steroids* **8,** 461 (1966).

397. E. Chang and V. K. Jain, *J. Med. Chem.* **9,** 433 (1966).

398. Z. Cekan and B. Pelc, *Steroids* **8,** 209 (1966).

399. Z. Cekan, Z. Vesely, J. Fajkos, and F. Sorm, *Steroids* **5,** 113 (1965).

400. Z. Cekan and I. Bartosek, *Steroids* **10,** 75 (1967).

401. H. L. Krüskemper and H. Breuer, *Verhandl. Deut. Ges. Inn. Med.* **67,** 387 (1961).

402. D. Gould, L. Finckenor, E. B. Hershberg, J. Cassidy, and P. L. Perlman, *J. Am. Chem. Soc.* **79,** 4472 (1957).

403. R. Horton and J. F. Tait, *J. Clin. Endocrinol. Metab.* **27,** 79 (1967).

404. P. Robel, R. Emiliozzi, and E. E. Baulieu, *J. Clin. Endocrinol. Metab.* **27,** 1290 (1967).

405. J. M. Saez and C. J. Migeon, *Steroids* **10,** 441 (1967).

406. U. Laschet, H. Nierman, L. Laschet, and H. F. Paarmann, *Acta Endocrinol.* Suppl. 119, 55 (1967).

407. R. Wiechert, H. Steinbeck, W. Elger, and F. Neumann, *Arzneimittel-Forsch.* **17,** 1103 (1967).

AUTHOR INDEX

Numbers in parentheses are reference numbers and are included to aid the reader in locating the citations in the text. Numbers in italic show the page on which the complete reference is listed.

A

Ackerman, J. H., 24(11), 40(204), 89(11), 217(11), 224(11), 230(93), 246(11), 247(11), 248(11), 279(11), *297, 299, 302*
Adelstein, G. W., 210(89), 215(89), 268(89), 275(88), *299*
Adezati, L., 84(262), *304*
Albert, A., 1(1), *297*
Alexander, J., 16(177), *302*
Allinger, J., 13(374), *307*
Allinger, N. L., 7(332), *306*
Alvarez, F., 216(58), 218(58), 219(58), *298*
Amerio, A., 84(261), *304*
Angyal, S. J., 7(332), *306*
Antoniades, H. N., 3(235), *303*
Antonini, F. M., 89(285), *305*
Ariens, E. J., 1(2), 91(2), *297*
Arnold, A., 24(11), 31(44, 184), 40(203, 205), 60(124), 88(63, 124, 203), 89(11, 49, 63, 287), 116(44), 138(124), 159(124), 165(63), 211(44), 217(11, 49), 224(11, 63), 246(11), 247(11), 248(11), 249(124), 251(124), 262(44), 263(63), 279(11), 280 (130), *297, 298, 300, 302, 305*
Arth, G. E., 144(136), 192(132), *300*
Atwater, N. W., 114(33), 115(33), 218(33), 226(33), 265(33), 272(33), *298*

B

Babcock, J. C., 78(269), 114(25), 119(18), 135(25), 136(25), 141(18), 146(156), 264 (73), *297, 298, 299, 301, 304*
Baeder, D. H., 174(90), *299*

Bagett, B., 14(225), *303*
Balant, C. P., 147(231), 148(231), *303*
Baldratti, G., 80(32), 83(244, 259), 84(243), 85(243), 113(32), 114(32), 148(32, 243), 211(32), 263(32), 264(32), 265(32), 281 (244), *298, 303, 304*
Bardin, C. W., 6(305, 323), 9(305), *305, 306*
Barnes, L. E., 82(65), 262(65), 281(65), *298*
Barnikol-Oettler, K., 138(110), *300*
Barton, D. H. R., 7(331, 340), 9(331, 346, 347), *306*
Bartosek, I., 17(400), *308*
Batres, E., 18(232), 59(34), 60(34), 80(23), 114(34), 137(161), 146(161), 157(34), 158(39), 163(23), 196(161), 197(161), *297, 298, 301, 303*
Baulieu, E. E., 5(385), 6(169, 311, 385), 8(169), 11(307, 311, 312, 385), 12(307, 308, 308a, 312, 404), 15(311, 312), 16 312), *301, 305, 308*
Beer, C. T., 5(319), *306*
Bein, H. J., 85(264), *304*
Bekaert, J., 84(260), *304*
Bendas, H., 37(16), 98(16), *297*
Berger, E., 83(251), *304*
Bergström, S., 13(316), *306*
Berköz, B., 159(57), 196(57), 215(57), 218(57), *298*
Bernhart, F. W., 90(290). 273(290), *305*
Bernstein, S., 246(15), *297*
Bertazzoli, C., 84(243), 85(243), 148(243), *303*
Bertolotti, E., 89(284), *304*

309

SUBJECT INDEX

A

1α-Acetylthio substitution, effect on activity, 24

17α-Alkyl substituent, role of 72–73

17α-Alkyl substitution, effect on activity, 21, 24

17α-Alkyl-testosterone, metabolism of, 16

17α-Alkylation
   effect on liver function, 19
   on oral activation, *see* Oral activation

Anabolic activity
   assay of, 31
   effect of 17α-alkyl substitution, *see* 17α-Alkylation
   of ring-A unsaturation, *see* Ring-A unsaturation
   role of β-sided surface, *see* β-Sided surface
   of upper surface, *see* Upper surface

Anabolic agents, general aspects, 1–2

Anabolic/androgenic effects, dissociation of, 90–91

Anabolic/androgenic ratio
   definition, 31
   of various steroids, 77–91

Anabolic effect, 1

Anabolic steroids
   clinical application of, 2
   definition of, 2
   effects on liver function, *see* Liver function
   tion

Androgenic activity
   effect of 17α-alkyl substitution, *see* 17α-Alkyl substitution
   of electronegative substituents, *see* Electronegative substituents

   of 19β-methyl group removal, *see* 19β-Methyl group removal
   of ring-A modifications, *see* Ring-A modifications
   of ring-A unsaturation, *see* Ring-A unsaturation
   α-face adsorption theory, *see* α-Face adsorption theory
   role of β-sided surface, *see* β-Sided surface
   of upper surface, *see* Upper surface

Androgenicity
   assay of, 31
   basic structure, 33

Androgens
   aromatization of, *see* Aromatization of androgens
   biosynthesis of, 3
   definition of, 1
   general aspects, 1–2
   production in man, 2–4

Androisoxazole, *see* 17α-Methyl-17β-hydroxy-5α-androstan[3,2-c]isoxazole

Androsta-1,4-dien-17β-ol-3-one 17-(cyclopent-1'-enyl) ether, 85

Androstanazol, *see* 17α-Methyl-17β-hydroxy-5α-androstano[3,2-c]pyrazole

5α-Androstane, androgenicity of, 33

Androstan-17β-ol, androgenicity of, 34

Androstanolon, *see* 17β-Hydroxy-5α-androstan-3-one

Androstenedione
   blood production rate, 9
   conversion to testosterone, 8
   plasma pool, *see* Plasma androstenedione pool
   secretion of, 9

Androsterone, in urine, 3, 9

321